fat

fat

It's Not What You Think

Connie Leas

Prometheus Books

59 John Glenn Drive
Amherst, New York 14228–2119

Published 2008 by Prometheus Books

Inquiries should be addressed to
Prometheus Books
59 John Glenn Drive
Amherst, New York 14228–2119
VOICE: 716–691–0133, ext. 210
FAX: 716–691–0137
WWW.PROMETHEUSBOOKS.COM

12 11 10 09 08   5 4 3 2 1

Library of Congress Cataloging-in-Publication Data

Leas, Connie.
    Fat : it's not what you think / Connie Leas ; foreword by George Mann.
       p. cm.
    Includes bibliographical references and index.
    ISBN 978–1–59102–612–9 (alk. pbk.)
    1. Fat—Health aspects. 2. Fatty acids in human nutrition. 3. Lipids in human nutrition. I. Title.

QP752.F3L43 2008
612.3'97—dc22

                                                              2008005597

Printed in the United States of America on acid-free paper

For my husband, Speed,
who, though his interest in the workings of the human body
approaches zero, is my biggest booster.

CONTENTS

CHAPTER 2. WHAT IS FAT, ANYWAY?   31

CHAPTER 3. HOW YOUR BODY DIGESTS AND USES FAT   37

CHAPTER 4. WHAT YOUR FAT CELLS DO   45

CHAPTER 5. WHO SAYS YOU'RE TOO FAT? 55

CHAPTER 6. WHY YOUR BODY WANTS TO KEEP ITS SHAPE 67

CHAPTER 10. WHAT'S WRONG WITH TRANS FATS

CHAPTER 11. THE PROBLEM WITH LOW-FAT DIETS

FOREWORD

This is an intriguing book because it deals with the cause of coronary heart disease and stroke. The author reviews many research projects and conjectures. Citizens should know that it is now about even money that each of them will die of a heart attack, the most frequent cause of death in the Western world. For fifty years the public has been told by officials of the American Heart Association and the National Heart Institute that this epidemic disease is caused by dietary saturated fatty acids and cholesterol. That advice is quite wrong. It is the greatest biomedical error of the twentieth century. The advice lingers, for selfish personal reasons and commercial avarice.

If not that—then what? In this detailed and scholarly book Connie Leas examines a welter of evidence, some science, some journalistic opinions. The accounts range from a recipe for mayonnaise to the probably true cause of our epidemic, the dietary trans fatty acids that were added to our food supply early in the twentieth century. While some may find this account of fat chemistry and physiology a little complicated, the explanations are relevant to an understanding of the issues.

A vast literature has grown up around this "heart problem." Citizens need to know the elements of this if they are to make the best food choices to prevent heart attacks and stroke. Readers will be appalled at the ways they have been misled in these matters. They need to know more about the trans fatty acids added to food some seventy-five years ago. They should know that exercise can protect them from unavoidable dietary trans fatty acids because muscles can use these materials for fuel. For those with hereditary hypercholesteremia, the dangerous statin drugs should be used with great care until safer and more effective drugs are found.

This is a long and complicated read, but worth the effort because it examines our primordial health problem and it explains a way for individuals to practice personal preventive medicine by making good food choices, and that is the best kind of preventive medicine.

George V. Mann, MD
Nashville, Tennessee

ACKNOWLEDGMENTS

I'm grateful to Laura Dolson, friend and About.com writer, for her enduring interest, thoughtful opinions, and hot tips throughout the many years of this book's gestation; to Dr. Bruce German, with the Department of Food Science and Technology at University of California at Davis, whose encouragement led me to believe in the value of my project; to Cora Morgan, Dr. German's assistant, who carefully reviewed and corrected an early version of the manuscript and pointed me to important new information; to Diane Hoxeng, friend and neighbor, for creating the illustrations; to Dr. George Mann, for reviewing the manuscript and writing the foreword, and for Melanie Gold, copy editor at Prometheus Books, for tidying up my words.

INTRODUCTION

This book started out as a sort of encyclopedic "everything you always wanted to know about fat. . . ." But as my research progressed, I began to encounter controversy, especially where saturated fat and cholesterol are concerned. The more I researched, the more convinced I became that we've been misled. For example, in the media, the term "saturated fat" is frequently preceded by "artery-clogging," which makes us form a mental picture of fat slithering down our gullets and working its way into our arteries, there to stop them up like a clogged sewer line. This is a sort of cartoonish description of what is known as the "lipid hypothesis" or "diet-heart" dogma. This idea, familiar to us all, holds that eating saturated fat and cholesterol leads to arterial plaques and heart disease. I quickly learned that plenty of scientists disagree with that notion (not to mention Julia Child, a butter lover who lived to be ninety-two). In fact, those who still adhere to the notion seem to be fading away. Dr. Uffe Ravnskov tells us "huge numbers of published medical studies reveal results that are totally at odds with this idea." Nevertheless, the idea persists, no doubt with the help of those who produce cholesterol-lowering drugs, low-fat products, and butter substitutes.

The idea also persists because of a phenomenon social scientists call an "informational cascade," which simply means that people—even scientists—tend to follow along with and propagate the ideas of someone who acts like an authority. As explained by John Tierney, in a *New York Times* article, "Diet and Fat: A Severe Case of Mistaken Consensus" (October 9, 2007), the informational cascade begins when a person of stature makes a public declaration—in this case, that fatty foods shorten your life. "If the second person isn't sure of the answer, he's liable to go along with the first person's guess. By then, even if the third person suspects another answer is right, she's more liable to go along just because she assumes the first two together know more than she does." One person after another assumes the rest can't all be wrong. In the case of the diet-heart hypothesis, the first person was, unfortunately, Surgeon General C. Everett Koop, who, in 1968, took his cue from Ancel Keys, whose erroneous but popular anti-fat message started the whole anti-fat campaign. Even after scientists were unable to confirm the diet-heart hypothesis, the mistaken consensus held fast, partly because of another sociological phenomenon called a "reputational cascade." This phenomenon kicks in when scientists fear that questioning the popular wisdom may pose a risk to their careers.

As Tierney explains, "Cascades are especially common in medicine as doctors take their cues from others, leading them to overdiagnose some faddish ailments (called bandwagon diseases) and overprescribe certain treatments (like the tonsillectomies once popular for children). Unable to keep up with the volume of research, doctors look to guidance from an expert—or at least someone who sounds confident." So what you have—and continue to have—is science by consensus. Indeed, when asked why the American Heart Association still warns against saturated fat, Dr. Ronald Krauss, who has headed up the Nutrition Committee of the American Heart Association and who stated in an interview that "the message of low fat has not been based on the best science," answered that the guidelines are made by voting. In other words, the bases for "authoritative" nutritional guidelines are closer to a Gallup poll than scientific studies.

It didn't take me long to come down on the side of those who deny that saturated fat and cholesterol are the culprits in heart disease—scientists such as Drs. George V. Mann, Mary Enig, and Uffe Ravnskov, and science writer Gary Taubes. This is partly because their arguments are the most compelling. But it's also because my instincts pull me in their direction. I tend to lean toward the commonsensical, which, in this case, favors natural processes over human-made interventions. Our bodies make cholesterol for a reason; saturated fat is a natural substance that has a rightful place in our diets.

Saturated fat, by the way, is saturated with hydrogen atoms—not glop as we might imagine. We imagine glop because of the way "saturated" is presented in the media—that is, preceded by "artery-clogging." But the word "saturated," like "triglycerides," "LDL," and other heart-disease-related words, is a scientific term. Scientific terms are tossed about frequently and casually in the media, often to instill fear and motivate us to purchase products. The terms have become so commonplace we think we know what they mean, but usually we don't. To help you become a savvy consumer of food and pharmaceuticals, I've devoted space to explaining some of the terms, trying to be as thorough as necessary without including too much mind-numbing detail. As Einstein said, "Make everything as simple as possible, but not more so."

Don't get the idea that this book is only about saturated fat and cholesterol. It's still intended to provide everything you need to know about fat. For this reason, the subjects are wide-ranging and cover body fat as well as food fat. Because the amount of material on these topics is overwhelming, I spent a lot of time sorting through it all and selecting what I consider to be the pieces that are the most important, interesting, and to the point. But whole books have been written on the subjects to which I devoted just a chapter—or even a section. For example, there's Gina Kolata's *Rethinking Thin*, which covers the genetics of obesity; Paul Campos's *The Obesity Myth*, which looks skeptically at conventional notions of weight; Ellen Ruppel Shell's *The Hungry Gene*, which explains the genetic components of appetite;

Uffe Ravnskov's *The Cholesterol Myths*, which exposes "the fallacy that saturated fat and cholesterol cause heart disease"; Gary Taubes's *Good Calories, Bad Calories*, which meticulously documents the history of the anti-fat movement and explains the role of insulin in obesity; and Nina Planck's *Real Food*, which promotes the health-giving properties of saturated fats. These books and others explore in depth some of the topics I discuss more briefly. The history behind the anti-fat movement is a case in point. It's a rather long and involved story, and I didn't want to get bogged down with it, so I gave it rather short shrift. But you can learn about it in both Ravnskov's and Taubes's books as well as in Taubes's excellent article, "The Soft Science of Dietary Fat," published in the journal *Science* and downloadable from www.sciencemag.org. I wanted this book to be more of a romp than a slog. I hope it strikes you that way.

Chapter 1.

GOOD THINGS ABOUT FAT

The fat you carry around has useful functions. It stores energy for future use, produces important chemicals, builds cell membranes and neural structures, provides padding, insulates against cold, supplies fuel, and supports your immune system. Fat can be your friend!

STORED ENERGY

The most obvious purpose of fat is for storing energy—something it does very well. Each gram of fat stores about nine calories' worth of energy, about twice more than a gram of carbohydrate. (Calories are the number of heat units you get by "burning" fuel.) One reason fat can store more energy than carbohydrates—gram for gram—is that it contains no water. Stored carbohydrates, on the other hand, are about two-thirds water. If, like plants, you stored carbohydrates for energy instead of fat, you'd be too huge to get around. Think of potatoes.

A single pound of fat contains about 3,200 calories of energy—enough to get you through many days without food. If you're a typical woman, your body is about 25 percent fat; if you're a man, it's more like 15 percent. Because you're probably lugging around many pounds of fat, you can go for weeks without food. If you're an average-sized man, for example, you might be storing 141,000 calories' worth of energy in your fat tissue, enough calories, at 2,000 a day, to last seventy days. The more stored fat, the longer you last without food. (Because women are fatter than men, in times of famine they outlast men.) One man who weighed 456 pounds went for 382 days without food, under medical scrutiny, and lost 276 pounds. (He's listed in the 1971 *Guinness Book of World Records*.) Don't try this at home, but if you find yourself a contestant on the *Survivor* show, you won't have to worry about starving to death.

HORMONE PRODUCTION

Energy storage is only one way your body uses fat. Your fat tissue also functions as a tremendously dynamic endocrine organ—the biggest endocrine organ in your body. (An endocrine organ is one that secretes hormones into the blood; hormones are substances, such as estrogen, that regulate various bodily functions.) Scientists have only recently discovered that fat tissue is not just a passive storage depot. Instead, fat cells are busily releasing hormones into the bloodstream just as the pituitary and thyroid do. So far, scientists have discovered roughly twenty-five different hormones and other compounds produced by fat cells. The hormones released by fat cells affect metabolism, weight, and overall health. Besides producing hormones, fat cells also release other compounds, including immune-system cells. Unlike other endocrine organs, which stay about the same size, fat can make more of itself. The more fat tissue, the more hormones get made—a situation that can be problematic, as I'll discuss in chapter 4.

CELL MEMBRANE MATERIAL

Fat is the primary building material both for cell walls and for the sheaths that surround and protect nerves. Because fat doesn't dissolve in water, it's the only material that's just right for this job (you wouldn't want your cell walls dissolving into mush). Acting as both a border and a wall, it defends and defines the cell's borders, keeping in those things that belong and excluding those that don't. But what belongs in and what belongs out constantly changes, depending on the cell's requirements. It's up to fat to regulate the traffic.

Fat can cope with the changes because it comes in hundreds of variations. In fact, the membrane of a single living cell can contain many types of fat, a quality that's critical for proper cell functioning. Cell walls are studded with proteins that function both as channels for letting chemicals in and out and also as sensors that trigger changes inside the cell. A wide diversity of fats make it possible for particular fats to interact with particular proteins, controlling their assembly and helping them interact with their environment. Some fats are also involved in a hugely complex web of signaling responses involving hundreds of different proteins. The signaling controls just about every cellular activity, including cell growth and movement, programmed cell death, and the transport of chemicals into and out of cells. Thus, cell membrane fats are key mediators of the cell's interactions with its environment.

Clearly, the cell membrane is not just a passive wall to stop cell innards from leaking away. Rather, it's a complex and dynamic cellular organ in its own right. The successful reception and processing of chemical messages depend on the health and integrity of cell membranes, of which fat is a key element.

PADDING

The spongy quality of fat makes it a good shock absorber. It protects our joints, internal organs, and eyeballs. The fat your body uses for padding is called structural fat and serves as a kind of bubble wrap to protect brittle tissues and vital organs from the bumps and bangs of everyday life. Structural fat doesn't expand or shrink after you reach maturity and doesn't store fat for energy—that is, it's not retrievable.

Structural fat is arranged in various configurations throughout your body to perform specific functions. The most familiar locations of structural fats are the soft pads under the tips of your fingers and toes as well as the pads of your feet. In the case of feet, the fat is encased between bands of collagen that are tightly bound to both the skin and the heel of the bone. The fat absorbs the impact between the heel and the ground. As Caroline Pond explains it in *The Fats of Life*, "[W]ithin milliseconds after impact, the bands recoil to their original shape." In this way the fat protects your bone and its joints from the shock of impact at every stride, and the recoil may help place your bones in correct position as you transfer your body weight from the heel to the ball of your foot. What's more, the mechanical properties of the structural fat in your foot provide feedback in the form of neural sensations that inform your nervous system about the position of your joints and how much your tissues are stretched. In this way, your body can automatically control stride, movement, and balance.

Structural fat situated behind and around your eyeball determines how your eyeball is positioned in your skull. Fat also encloses the large optic nerve. In the lungs, specialized fats function as a sort of anti-glue, enabling the tiny pockets in the lung to fill easily with air and preventing them from sticking together. The pockets pack into a small space when the lungs are deflated, but respread readily on inflation.

INSULATION

Fat located below the surface of your skin provides some insulation from the cold, although fur or feathers would do a better job of it. Fat limits the transfer of heat from your body's core. Because blood flows through muscle at a higher rate than it does through fat, especially during vigorous exercise, your body loses heat more quickly from superficial muscles than from superficial fat. But the main evolutionary purpose of fat is for survival during periods of food scarcity, not as an insulating mechanism.

THE IMMUNE SYSTEM

The energy you store as fat can bolster immunity and combat infection. When you've got an infection or a severe wound or burn, the concentration of fatty acids in your bloodstream rises just as it does if you've gone for a long period without eating. This is because your immune system requires fat to "feed" your infection-fighting lymphatic system. Specifically, it needs polyunsaturated fatty acids to build the cell walls of lymphocytes (white blood cells) that are your first line of defense against disease. In fact, nearly all major lymph nodes and parts of their connecting ducts are embedded in fat tissue.

Skinny people are more likely to succumb to infectious disease than fatter people. In fact, people rarely die just from starvation. Instead, their loss of body fat lowers their resistance to disease so effectively that the immediate cause of disease is usually an infectious disease or pneumonia-like inflammation of the lungs. Apparently, the immunity conferred by fat cells is a result of the anti-inflammatory substances produced by the fat cells.

Using rodents as test subjects, a team of researchers at Indiana University, Ohio State University, and Johns Hopkins University discovered that sudden removal of fat (from the groin area) decreased the animals' immune systems. They found that even a small decrease in body fat decreases immunity, which in turn increases susceptibility to disease. The researchers also found that immune system function improved after the fat tissue grew back.

PROTECTION FROM TOXIC SUBSTANCES

Body fat can also mediate the effects of environmental toxins on your nervous system. Many pesticides and other toxic substances are fat soluble. If you are exposed to such chemicals and have little body fat to dilute them, they are likely to concentrate in nervous tissue where they can lead to neurogenerative disease. Dr. Andrew Weil, in his book *Healthy Aging*, even postulates that "this is the reason that ALS (amyotrophic lateral sclerosis—Lou Gehrig's disease) appears more frequently in athletes. At the Integrative Medicine Clinic at the University of Arizona, I have seen a number of men in their thirties with this devastating diagnosis, an unusually young age group for the onset of ALS. All had a history of participating in extreme sports and competitive, ultra-athletic events, and all were extraordinarily lean."

Dr. Roy Walford, a pathologist at the University of California, Los Angeles, was a leading authority on the concept of using a restricted-calorie diet to combat the effects of aging and disease. Apparently, being on a near-starvation diet tells your body that food is scarce. This is a stressful situation that triggers a gene that in turn triggers physiological changes that slow down the aging process in the interests of survival. Because a calorie-restricted diet has been shown to prolong the lives of laboratory animals, Dr. Walford followed such a regimen for the last thirty years of his life. He died of Lou Gehrig's disease at age seventy-nine. He was five foot nine and weighed 130 pounds. While Dr. Weil's theory about low body fat and Lou Gehrig's disease may not stand up to scientific scrutiny, it seems like a plausible notion. It also illustrates how little we understand about ideal body composition.

SEXUAL ATTRACTION

A man with a paunch used to be considered good marriage material. His paunch showed that he was well fed, and thus well off. Perhaps more than that, his paunch may have also been a sign of his virility: It's the male sex hormone, testosterone, that's responsible for this selective buildup of fat in his abdominal region. Among primates, such as apes, it's the dominant males that sport the prominent paunches.

For women, plump hips and thighs were probably the best man bait. Because a wide pelvis is best adapted for childbirth, accentuating the pelvis with fat hips and thighs advertised a potentially successful breeder. It's less clear whether large breasts had much attraction for the prehistoric man, since there's no special relationship between breast size and capacity for milk (but maybe the guys didn't know that).

Noting that, on average, an adult woman has about twice as much body fat as the average man, Desmond Morris, in his book *The Naked Woman, A Study of the Female*, theorizes that female fatness is one of the traits—along with a higher voice and finely boned face—that evolved via natural selection. Fatter—and therefore more childlike— women were more successful at mating, he believes, because those infantile traits elicited protective responses in males.

FAT-FREE PEOPLE

A rare disorder, called lipodystrophy, strips the body of fat, in either just a few areas or everywhere. The condition comes in various forms. Sometimes it's caused by defects in the genes responsible for creating the enzymes needed to manufacture fat or maintain fat cells. Infants born with this genetic defect show muscle definition instead of the normal baby fat. They usually have voracious appetites and grow tall but never deposit an ounce of fat on their frames. Other forms of the disease, which appear to be unrelated to gene mutations, show up in childhood or adolescence. In some cases, fat disappears from the entire body; in other cases it disappears only from the upper half, with the lower half sometimes overly fat. In these cases, the cause is thought to be an immune system malfunction, in which the body attacks and destroys its own fat cells.

Studying people with lipodystrophy points out the importance of fat for normal body functioning. People with lipodystrophy are plagued by metabolic problems. In particular, their bodies are highly resistant to insulin. As a result, they develop diabetes. Without fat cells to store the fat, the fat in the bloodstream—whether extracted from digested food or synthesized by the body—has nowhere to go, so it either stays in the blood or is deposited in tissues like the liver, where it is converted to glycogen, then released later into the blood as glucose. With all the excess fat and sugar in their bloodstreams, people with lipodystrophy are at just as high a risk for heart disease as those with diabetes. Lipodystrophy is common among people with AIDS. In this case, fat disappears from the face, arms, and legs, but accumulates in the neck, trunk, and abdomen. Scientists don't know what causes lipodystrophy in people with HIV.

The functions of fat mentioned here don't begin to cover all its uses, partly because so little is known. Scientists have only recently begun to delve seriously into its biological complexities, spurred partly by the discovery of leptin in the mid-1990s (see chapter 6) and partly by the drive to find pharmaceutical solutions to the "obesity epidemic." For sure, our bodies are unimaginably complex and intricate systems, and fat is a key, though little-understood, player.

Chapter 2.

WHAT IS FAT, ANYWAY?

†he fat you eat and the fat on your body are basically the same—mainly a substance called "triglyceride." (The other two main classes of fats are phospholipids, such as lecithin, and sterols, such as cholesterol—which is technically an alcohol. At any rate, because these substances comprise only a small percentage of our fat intake, I'll limit this discussion to triglycerides.) You've probably come across the term in magazine and newspaper articles. Perhaps you've had blood work done and received a report that includes your triglyceride level. In any case, it's just fat.

To help you understand fat in its various forms (saturated, monounsaturated, and polyunsaturated) you need to delve into a little chemistry—but just a little. It's the only way to make sense of this.

TRIGLYCERIDES DECONSTRUCTED

Triglycerides are so named because they are constructed of three (tri) chains of fatty acids bonded at three positions to an alcohol (glycerol) backbone. The glycerol backbone is an alcohol, similar to ethanol, the main active ingredient of alcoholic drinks. Here is a schematic of a triglyceride molecule (molecules are made of atoms that bind to one another by sharing electrons—atomic particles with negative charges).

Structure of a triglyceride.

FATTY ACIDS DECONSTRUCTED

The fatty acid portion of the triglyceride molecule consists of a chain of carbon atoms, each of which is attached to at least one hydrogen atom. Here's an illustration of a triglyceride molecule showing three fatty acids connected to the glycerol backbone:

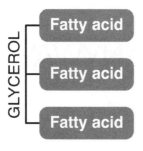

The illustration is just one example. Scientists have discovered dozens of different fatty acids. (Fatty acids are called "acid" because the hydrogen atom at the end easily dissociates from the rest of the molecule to form a hydrogen ion, the hallmark of an acid.)

Unless a triglyceride has been tampered with, it will be a mixture of different fatty acid carbon chains or "tails." In other words, in nature, you won't find triglycerides composed of three identical fatty acids. Each fatty acid has a name, such as stearic, oleic, and linoleic. A triglyceride molecule could contain one of each.

The fatty acids making up the triglyceride molecule differ in a number of ways: the length of the carbon chain, how the carbon atoms are connected to one another, the arrangement and number of hydrogen atoms on the carbon chains, and whether the chains are straight or curved. It's these variations that give the fats we eat distinctive characteristics with regard to taste, texture, melting temperature, digestibility, and other properties. And it's the particular atomic configuration of the fatty acid that determines whether it is considered saturated, monounsaturated, or polyunsaturated. Specifically, the designation of a fat into one of these three types depends entirely on the existence of double bonds between the carbon atoms, as I'll explain below.

SATURATED, MONOUNSATURATED, AND POLYUNSATURATED FATTY ACIDS

You are doubtless acquainted with the terms "saturated," "monounsaturated," and "polyunsaturated." But you may not understand how one type differs from another, a situation I hope to remedy in the following discussion.

The Basics

In the fatty acid molecule, all the carbon atoms in the central portion of the carbon chain have room for two hydrogen atoms, as shown here:

$$\begin{array}{ccc} H & H & H \\ | & | & | \\ C{-}C{-}C \\ | & | & | \\ H & H & H \end{array}$$

But not all fatty acids have two hydrogens for each carbon atom. Sometimes, a carbon has just one hydrogen atom attached to it, in which case it makes up for that lack by forming a *double bond* (C=C) with the carbon atom next to it. (A "bond" is the force—in this case the shared electrons—that holds atoms together; carbon is one of the few elements that will join together by sharing more than one electron.) The difference between saturated, monounsaturated, and polyunsaturated has to do with whether or not double bonds exist and, if so, how many.

Saturated fatty acid

The molecules in a saturated fatty acid contain *no* double bonds. That is, the carbon atoms all are "saturated" with hydrogen atoms—two hydrogen atoms for each carbon atom (the last carbon in the chain has three hydrogens):

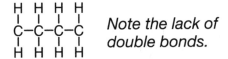

Note the lack of double bonds.

Saturated fat (fragment of chain).

Saturated fatty acids are all straight chains. As such, they pack neatly together to form solid rather than liquid material, giving them the designation of fats, rather than oils.

Monounsaturated fatty acid

The molecules that make up monounsaturated fatty acids contain just *one* double bond. That is, at one point along the carbon chain, two hydrogen atoms are missing from the carbon atoms. To make up for that lack, carbon atoms with the missing hydrogen atoms form a bond with the carbon atom next to it. But only one set of carbon atoms in that molecule has the double bond, hence the term "*mono*unsaturated."

$$
\begin{array}{cccc}
\text{H} & \text{H} & \text{H} & \text{H} \\
| & | & | & | \\
\text{C} & \text{--C} & \text{=C--} & \text{C} \\
| & & & | \\
\text{H} & & & \text{H}
\end{array}
$$

Note the single double bond. (=).

Monounsaturated fat (fragment of chain).

Monounsaturated fatty acid chains are bent. Thus, a more realistic image of a monounsaturated fatty acid (in this case oleic acid, the predominate fatty acid of olive oil) is this:

As you can see, such a configuration wouldn't stack tightly together. For this reason, monounsaturated fatty acids are liquid at room temperature and are classified as oils.

Polyunsaturated fatty acid

The fatty acid chains that make up polyunsaturated fat contain two or more double bonds between carbon atoms in the chain:

$$
\begin{array}{ccccccc}
\text{H} & \text{H} & \text{H} & \text{H} & \text{H} & \text{H} & \text{H} \\
| & | & | & | & | & | & | \\
\text{C} & \text{--C} & \text{=C--} & \text{C} & \text{--C} & \text{=C--} & \text{C} \\
| & & | & & | & & | \\
\text{H} & & \text{H} & & \text{H} & &
\end{array}
$$

Note multiple double bonds.

Polyunsaturated fat (fragment of chain)

Polyunsaturated fats are shaped like a stick with a double bend, which makes them liquid at room temperature and thus classified as oils.

A SEPARATE CASE: *TRANS* FATTY ACIDS

Trans fats are a special case. They are manufactured from polyunsaturated fats and processed in such a way that the normally bent structure of the polyunsaturated fats are straightened out, giving them the same physical properties as animal fats, such as butter. Because trans fats are inexpensive and unnaturally solid, they are widely used by the processed food industry. They have a long shelf life and provide the consistency needed for spreads and shortenings (and Oreo cookie filling). Chapter 10 explains more about trans fats and why they're not good for you.

WHY ALL THE FUSS ABOUT SATURATED FATS?

As you can see, saturated fat isn't saturated with cholesterol. It's saturated with hydrogen—the same element in water (H_2O). What's more, the triglycerides you eat and those your body uses are made up of mixtures of monounsaturated, polyunsaturated, and saturated fat molecules. For example, palm oil, which we think of as saturated fat, contains less than half saturated fat (called palmitic fatty acid), with the remaining comprised of polyunsaturated and monounsaturated fatty acids. Olive oil, which is primarily oleic acid, a monounsaturated fat, contains 14 percent palmitic fatty acid. Nature provides us with mixtures and our bodies use mixtures. As long as you stick with what nature provides, there's no need to fuss about fats.

Chapter 3.

HOW YOUR BODY DIGESTS AND USES FAT

So you've been eating fat. Now what? First, it has to be digested. Digestion, as you probably know, begins in your stomach. Because fats don't dissolve in water, they're slower to digest than proteins and carbohydrates.

FIRST, THE STOMACH

The first stage of digestion is mechanical. Powerful muscles in your stomach wall churn the food, breaking fats into tiny droplets—a process called emulsification. Next, your stomach produces an enzyme called "gastric lipase" that breaks fat into its component parts. Specifically, gastric lipase breaks the bonds—called "ester bonds"—that link the fatty acids to their glycerol backbones. This constitutes digestion. Only about 10 percent of fat digestion takes place in the stomach, although if you regularly eat a high-fat diet your stomach will learn to produce more gastric lipase to increase its fat-digesting capacity.

(Infants, by the way, can begin digesting fat in their mouths. Unlike adults, babies' saliva contains a lipase that breaks down fat in their mouths. Human milk also contains a lipase to help with fat break-down. Thus, infant stomachs have less digestive work to do than adult stomachs.)

NEXT, THE INTESTINES

After being partially digested in the stomach, food enters the small intestine. At this point, a hormone called cholecystokinin sends a message to the brain that it's time for food digesting substances to be released. The brain complies by sending signals to the gallbladder and pancreas, triggering the release of bile and enzymes. With the help of these substances, emulsification and digestion continue.

Bile, which is produced in the liver, facilitates the emulsification step. Bile enters your intestines by way of the gall bladder and bile duct. (The gall bladder is where bile is stored until it receives signals from hormones and nerves that it's needed.) The acids in bile act as a kind of detergent, breaking clumps of fat into small globules of emulsified fat. At this point, the fat's not digested yet; it's only broken into small globules. The small globules float about in the watery environment of the gut.

Next, pancreatic lipases, which are manufactured by the pancreas and released into the small intestine, continue breaking down the fats. (By now, you've probably figured out that a lipase is an enzyme that breaks down fat.) Fats, as explained in chapter 2, consist primarily of triglycerides: three fatty acids hooked to a glycerol backbone. Like gastric lipases, the pancreatic lipases break the ester bonds that link fatty acids to their glycerol backbone. (The lipase manufactured in the pancreas is slightly different from gastric lipase: each breaks down slightly different mixtures of fatty acids.)

After lipase breaks down the triglycerides into their component parts, those parts clump together into an aggregate consisting of bile salts, phospholipids, and cholesterol, which then diffuse into special cells of the intestinal wall.

GETTING THROUGH INTESTINAL WALLS

In the intestinal walls, special cells reassemble the components back into trigylcerides and create a new package, called a chylomicron, with a fatty inner core and protein outer core. The chylomicron package is the largest of the lipoproteins, a term you've probably heard before in the labels "high-density lipoprotein" (HDL) and "low-density lipoprotein" (LDL). Lipoprotein packaging makes it possible for fat to be transported in a liquid.

Actually, only long-chain fatty acids—which are the most common among food fats—go through the packaging processes. The less abundant short- and medium-chain fatty acids do not need to be repackaged by intestinal cells. Instead, these smaller compounds can move across the intestinal cells and directly into the blood. Here, they bind to proteins so they can move safely around your body. Because medium-chain fatty acids are so easily digested, healthcare professionals give foods that contain an abundance of these fats, such as coconut oil, to people with liver diseases that impair normal fat absorption.

TRANSPORTATION TO TISSUES

Lymph vessels pick up the chylomicrons and deposit them into the lymphatic system. The lymphatic system then ferries the packages into the blood, where they eventually make their way into the tiny blood vessels that permeate your various tissues, including fat and muscle. The tissues receiving the packages produce an enzyme called "lipoprotein lipase" that breaks up the packages into their constituent parts, making it possible for them to be absorbed into the tissue cells. Here they're reassembled into triglycerides and stored. What's left of the package, including cholesterol, goes to the liver for processing.

UNABSORBED FATS: THE LAXATIVE EFFECT

If all systems are working well and there's enough bile and various lipases to break down the fat, your body digests and absorbs the fat so efficiently that nearly all of the common dietary fats eventually arrive at the blood and lymphatic systems. But sometimes fat is not absorbed, either because of disease (poor absorption of fats is a common symptom of liver diseases), or because certain fats, such as castor oil, are constructed in a way that prevents lipase from breaking them down. This is why castor oil works as a laxative. It passes undigested into the bowels. Like castor oil, the "fat substitute" called Olestra is molecularly constructed such that lipases cannot break it down for absorption, and the fat passes undigested to the large intestine. If eaten in large quantities, Olestra can cause diarrhea.

SUBSTANCES ALONG FOR THE RIDE

At the time your body is digesting fats, fat-soluble substances, both good and bad, can slip in along with the fatty acids. Vitamins A, D, K, and E are absorbed in this way, but so are contaminants such as DDT (a pesticide), and PBDE (the flame retardant used to coat highly flammable materials such as plastic, foam, and some fabrics). These contaminants are too large and too insoluble in water to be excreted via the kidneys, so they accumulate in fat tissue. Contaminants may also accumulate in breast milk. Because human breast milk is 4 percent fat, toxins such as PBDEs in breast milk are passed from mothers to infants.

DIGESTIVE SPEED

Digestion and absorption of fats are relatively slow, much slower than those of all but the most complex carbohydrates. If you eat a meal that includes both fats and simple sugars, the simple sugars will usually have been digested, absorbed, and distributed to tissues before significant quantities of fat even get into the bloodstream. In fact, it may take more than six hours for your body to digest and absorb a meal that contained large amounts of fats. One reason for this is a reflex called the "ileal brake." The "brake," which is located in the last portion of the small intestine, stops the movement of partially digested food at the ileum. Thus, it prevents material from leaving the small intestine until the last of the fatty acids have been digested and absorbed through the walls of the small intestines.

Different types of fats vary in how quickly they are absorbed through the lining of the gut and into the blood. The most quickly broken down are those triglycerides that contain medium- and short-chain fatty acids, such as milk fats. The slowest to break down are fats composed of very long chain polyunsaturates, such as fish oils, which contain essential fatty acids. In other words, essential fatty acids are digested less efficiently than those fats whose main function is to produce energy. (See chapter 9 for a complete discussion of essential fatty acids.)

THE FOOD-TO-FAT CONVERSION: WHEN YOU EAT MORE THAN YOU NEED

Almost all tissues can use glucose as fuel. (Glucose is a six-carbon sugar that results from the digestion of carbohydrates.) Within minutes of eating food containing sugars or starch, the pancreas secretes insulin. In fact, the concentration of insulin in the blood rises by about five times that of its before-eating concentration. Insulin acts on receptors in cell membranes, prompting tissues to take up glucose from the blood and use it for energy production or storage.

If there's more glucose in the blood than your tissues need immediately, the liver and muscles take it up and convert it to glycogen, a type of long-chain sugar. The glycogen is stored until it's needed for energy. However, the muscles and liver can store only so much glycogen. Unlike plants, such as potatoes and carrots, that store glucose as starches, we are unable to store large quantities of starch. Any excess glucose that remains in the blood after glycogen deposits are filled up is converted into fat. That is, excess calories, regardless of source, are stored as triglycerides. In other words, even too many bean sprouts can be stored as fat.

THE FOOD-TO-ENERGY CONVERSION: WHEN YOU EAT LESS THAN YOU NEED

After about six hours of fasting, various tissues begin to deplete their reserves of glucose and glycogen and the liver switches from storing glucose to releasing it back into the blood. If the glycogen released by the liver (and muscles) is insufficient to supply fuel for the cells, the amount of insulin in the blood decreases. The drop in insulin stimulates the enzyme lipase to begin breaking stored fat into its component parts—fatty acids and glycerol—a process called "lipolysis." Lipolysis must occur before fats can be used as fuel. In addition to being stimulated by a drop in insulin, lipolysis is also enhanced by a rise in the hormone glucagon, produced by the pancreas, which stimulates the liver to release glycogen.

Another hormone, called noradrenalin, can also stimulate lipolysis. Under conditions of stress and excitement, nerves that permeate your fat cells stimulate your body to release noradrenalin. Lypolysis that is stimulated by noradrenalin also increases blood flow through fat tissue, making it possible for the newly released fatty acids and glycerol to be carried away and taken up by tissues elsewhere in the body. Some evidence indicates that the nervous system, depending on the situation, selects and stimulates certain specific fat cells to break down and release fatty acids and glycerol. For example, nervous stress may stimulate certain fat cells to break down fats for energy while brief exercise might stimulate different fat cells to do the same.

Within seconds of being released from fat cells, fatty acids find their way into the muscles, liver, and other fat-using tissues. In the cells of these tissues, the fatty acids are oxidized in about a dozen different biochemical steps, to produce ATP, carbon dioxide, and water. (ATP is adenosine triphosphate, a chemical that transfers the stored energy from broken chemical bonds for use in biological processes.) This process requires a good supply of oxygen.

FAT AS FUEL FOR SUSTAINED EXERCISE

Under normal circumstances, your muscle tissue doesn't produce as much lipoprotein lipase as fat cells do. (Remember, lipoprotein lipase is the enzyme that breaks down fats to use for energy.) Because muscle tissues normally use glucose and glycogen for energy, they don't break down triglycerides to use for energy. Instead, fat cells do most of this work. However, during periods of prolonged fasting and sustained exercise, supplies of glucose and glycogen may be in short supply. To compensate for this deficit, the muscle tissue (including heart muscle) begins producing more lipoprotein lipase. In this way, muscles can now burn whatever amount of fat it may be harboring. At the same time, fat cells begin producing less lipoprotein lipase. During periods of starvation, proteins—especially those that make up muscle—also begin to break down into the amino acids that are converted to glucose. Your body uses protein components for fuel long before fat reserves are exhausted. This is why dieters may find that muscle wasting and weakness may accompany their efforts to lose fat.

If a marathon runner uses glucose more rapidly than it can be replenished, glucose supplies to the brain can become insufficient and the athlete may become light-headed and collapse with exhaustion. To avoid early depletion of glucose, marathon runners typically practice "carbo loading"—forcing themselves to eat lots of starchy food during the few days before competing. By doing this, they increase their stores of glycogen in the liver and muscles to higher than normal levels. (After a large meal and if we're sedentary, stored glycogen may be sufficient to last us for about twenty-four hours. But most people normally keep such reserves well below the maximum capacity. While stored glycogen will supply about 1,900 calories, fat tissue can supply up to 130,000 calories, depending on how much fat tissue we have.)

Until recently, fats have been considered inconsequential or even detrimental to endurance exercise. It was assumed that it took too long to digest, transport, and oxidize fats to do the athlete much good. Now we know that highly trained athletes can oxidize fats for energy at a significantly higher rate than less fit people. In fact, several studies have demonstrated that athletes can improve their endurance by upping their fat consumption from 15 percent of their calorie intake to 32 percent to 55 percent. While carbohydrates should remain an important part of the athlete's diet to ensure that stores of glycogen are not compromised, experts are now recommending an increase in the proportion of fat the athlete eats. Specifically, sports medicine experts now recommend a diet composed of 20 percent protein, 30 percent carbohydrates, and 30 percent fat, with the remaining 20 percent of the calories distributed between carbohydrates and fat, depending on the intensity and duration of the sport—higher carbohydrates for intense sports of shorter duration.

Through its complex web of interconnected systems, your body efficiently disposes of the fat you eat—using it for energy, metabolizing it for cellular requirements, or storing it for later use. The cells and tissues that store the fat have their own complex systems, which I discuss in the next chapter.

Chapter 4.
WHAT YOUR FAT CELLS DO

excess calories that your body converts to fat are stored as triglycerides in your fat cells. The triglycerides sit there until your body either converts them into more complex fats or burns them for energy, depending on what's needed. But the fat cells do more than passively hold fat. Scientists have only recently learned that they also spew a variety of other substances—mostly hormones—into your bloodstream, which can create problems.

THE ANATOMY OF FAT CELLS

The technical term for fat cells is "adipocytes," from the Latin *adipatus*, which means greasy. A bunch of fat cells constitute adipose tissue. I'll just call it fat. Fat cells are roughly six-sided and fit together sort of like bubble wrap. Each cell contains a globule of fat, mostly consisting of triglycerides. Fat cells are huge compared with other body cells. Moreover, unlike cells in other body tissues, they can expand or contract, depending on what's needed. When you eat more than you need, the cells become very full, swelling to as much as six times their minimum size. A thin-looking person may have the same number of fat cells as a moderately obese person, but the cells of the obese person are greatly enlarged.

If the fat cells are filled to capacity, your body adds new ones. The overextended fat cells send a signal to nearby immature cells to start dividing. New cells are very tiny. It takes a million of them to store the calories in a Life Savers candy. Once they've been added, the cells tend to hang around. If you lose weight, they'll shrink and become less metabolically active, but the cells remain.

FAT CELL NUMBERS

Babies are born with about five billion fat cells—about one-fifth that of adults. Between infancy and adolescence, fat cells continue to be added. After adolescence, the addition of new cells should taper off, although new cells can continue to accumulate throughout life. Generally, if you're mildly obese, your existing fat cells simply become very full. However, if you're extremely obese or your obesity began in childhood, fat cells can increase substantially.

Adults of average weight have 25 to 40 billion fat cells; obese people can have as many as 100 to 200 billion. (Estimates vary.) In any case, people vary in the number of fat cells they carry. Some have ten times as many fat cells as others of similar body mass.

FAT AS A HORMONE PRODUCER

Until fairly recently, scientists believed that fat cells simply stored fat until it was needed. Now they're discovering that fat cells function collectively as an endocrine organ, busily releasing hormones into the bloodstream just as the pituitary and thyroid do. The hormones and other substances released by fat cells are potent: they affect metabolism, weight, and overall health. Unlike other endocrine organs, which stay about the same size, fat can make more of itself. The more fat tissue, the more hormones get made, which can be problematic. What's more, fat cells release other substances, including vasoconstrictors (chemicals that narrow blood vessels), macrophages (immune system cells), and blood clotting agents. Because larger fat cells are more metabolically active than small ones, they are more likely to churn out more harmful substances.

Roughly twenty-five different signaling compounds are produced by or stored in fat cells. For a summary of the key compounds, see appendix A.

HORMONES WORTH SPECIAL MENTION

Two hormones deserve special mention. One is leptin, which plays a key role in appetite. Leptin circulates in the blood in proportion to the amount of fat stored in the cells. As the amount of fat stored in cells increases, the amount of leptin in circulation also increases, which should suppress appetite. I'll explain more about how leptin works in chapter 6.

The other hormone is adiponectin, a *beneficial* hormone that stimulates your body to burn fatty acids as fuel, thus reducing the amount of fatty acids in your blood. (If your blood contains excessive fatty acids, the acids can get stored directly in the liver, heart, and muscle cells, damaging those cells. Also, a high level of fatty acids in the blood inhibits nitric oxide, a compound that helps relax blood vessel walls and lower blood pressure.) The more adiponectin your fat cells produce, the better. Unfortunately, the more fat you carry around, the *less* adiponectin your fat cells make. In obese people, then, adiponectin is repressed; if the obese person loses weight, adiponectin is increased—a good thing.

Whether your fat cells produce a lot or a little adiponectin affects whether you are more or less likely to suffer from heart disease: adiponectin reduces vascular inflammation, helps prevents atherosclerosis, and even seems to improve the balance of HDL and LDL cholesterol.

The amount of adiponectin your body produces also affects whether you are more or less likely to suffer from Type 2 diabetes. This is because adiponectin makes cells more sensitive to insulin, thus enabling them to efficiently take up glucose from the bloodstream. A study of Pima Indians showed that people with low levels of adiponectin were at significantly higher risk of developing Type 2 diabetes than those with relatively high concentrations of adiponectin; conversely those with high levels appeared to be protected from the disease.

Scientists are exploring a link between adiponectin and the metabolic syndrome—sometimes called syndrome X. The metabolic syndrome is a cluster of risk factors associated with heart disease and Type 2 diabetes. The risk factors include abdominal obesity, elevated blood triglyceride level, low HDL cholesterol, raised blood pressure, insulin resistance, tendency toward inflammation of blood vessels, and a tendency toward clot formation.

FAT DEPOTS (STORAGE AREAS)

As you're probably well aware, fat is not evenly distributed throughout your body. It's partitioned into several large and numerous small "depots" scattered around your body. Some of the fat, called subcutaneous fat, is located between the skin and superficial muscles. Some fat, called visceral fat, collects deep into the abdomen and nestles around the heart, intestines, spleen, kidneys, and reproductive organs. A small fraction of fat is embedded between skeletal muscles. This kind of fat is called intermuscular fat. Typically, intermuscular fat constitutes about 5 percent of total fat, but can be as high as 20 percent.

The most conspicuous depot is abdominal fat (the paunch), which is fat that collects around your navel. Abdominal fat can be composed of both subcutaneous fat and visceral fat. Abdominal fat arises from the midline on the outer wall of the abdomen and expands to the side as it thickens (the love handles). In men, paunches are usually somewhat thicker around the navel. In women, fat tends to collect below the navel and in the thighs and buttocks (also, women's breasts consist mainly of fat).

Scientists are now speculating that certain fat depots may hold rare "valuable" fatty acids to be used specifically for adjacent tissues that need them, rather than releasing them into the blood circulation for general consumption. Such ideas suggest that fat tissue may actively select and organize the fat it stores, rather than simply serve as a general repository.

APPLES, PEARS, AND OTHER FAT STORAGE ARRANGEMENTS

Two types of fat storage arrangements are commonly referred to as apple shape and pear shape. If you're apple shaped, you carry your extra fat in the abdominal area. If you're pear shaped, you carry your extra fat in the lower body—hips and thighs. Men tend to be apples and women tend to be pears, although women tend to redistribute fat toward apple shape after menopause.

While your genetic makeup controls your overall fat storage patterns, the distribution mechanism is the work of various hormones and an enzyme called "lipoprotein lipase" (LPL). As mentioned in chapter 3, any cells that use fatty acids for fuel or storage secrete LPL. LPL breaks fat into its component parts, making it possible for the cells in your body to use it as fuel or store it as fat. The more LPL activity in the cells, the more fatty acids it will absorb. In both individuals and populations, LPL activity is unevenly distributed. Women, for example, have more LPL activity in their fat cells than men. In men, the LPL activity is greater in the abdominal area; in women, it's greater in the hips and buttocks. LPL activity also changes over time. In menopause, a drop in estrogen shifts LPL activity to the abdominal region, which is why older women develop thick midsections. For men, decreasing testosterone increases LPL activity in the abdomen, with the attendant increase in abdominal fat.

For some people, different parts of the body have an exaggerated tendency to store fat, a condition known as *lipophilia*. You may have this condition yourself. You've certainly seen it in others, mostly in women who carry the extra fat in the hips and buttocks, and in men who carry it in their bellies. The condition, which is unaffected by dieting or exercise, is thought to be the result of excessive LPL activity in certain areas of the body. In other words, in certain areas, such as the hips or abdomen, the LPL enzyme is more aggressive at grabbing onto and absorbing fatty acids, with the result that excessive fat is stored in these areas.

VISCERAL FAT VERSUS SUBCUTANEOUS FAT

Where your fat is stored matters. Fat cells behave differently in different parts of the body. Fat stored around the organs in the abdomen (visceral fat) is more metabolically active than fat stored in the hips and thighs. Visceral fat is more dangerous than subcutaneous fat. It has a higher association with diabetes and high blood triglycerides than fat stored elsewhere. Visceral fat also produces more of the inflammatory and clot-promoting compounds than the subcutaneous fat that's distributed around other parts of your body. Moreover, the hormonal secretions of visceral fat go straight to the liver and may interfere with the liver's ability to help regulate both blood glucose and cholesterol. Visceral fat also secretes angiotensinogen, which makes angiotensin II, a powerful hormone that constricts blood vessels and also directs the kidneys to absorb more sodium, a condition linked to high blood pressure. Subcutaneous fat, on the other hand, is a bigger producer of leptin, the hormone that regulates appetite as well as blood pressure and reproductive functions.

VISCERAL FAT AND DIABETES

In one study of children, researchers looked at whether overall body fat or visceral fat had greater effects on risk factors for Type 2 diabetes. (They used children so they could rule out factors such as smoking, alcohol, stress, and sex steroids.) What they found was that overall body fat affects the cells' *sensitivity* to insulin. Visceral fat, on the other hand, affects the *amount* of insulin that's present in the blood during a fasting state. In both cases, then, excess fat had an influence on the risk for developing Type 2 diabetes: it decreases the sensitivity of cells to insulin, making them resistant, and also increases the amount of insulin produced by the pancreas.

One hypothesis to explain why visceral fat is linked to Type 2 diabetes is the fact that visceral fat breaks down into free fatty acids more easily than fat in other areas, resulting in a greater turnover in fatty acid releases. What's more, the fatty acids are released into the hepatic portal vein and go directly to the liver. Too many free fatty acids coming through the portal vein seem to make the liver produce too much glucose. With so much glucose, the body pumps out more insulin to try to control the sugar's high levels.

VISCERAL FAT AND EXERCISE

Physical inactivity can have a profound effect on the buildup of visceral fat. Luckily, for overweight, sedentary people, exercise can significantly reduce visceral fat in a fairly short time period. In his studies, in which he chose participants who were both overweight and sedentary, Dr. Cris Slentz, of the Duke University Medical Center, found that jogging for seventeen miles a week (or its exercise equivalent) resulted in significant declines in visceral fat as well as subcutaneous abdominal fat. In the study, some participants did not exercise at all; others exercised in varying amounts. None were allowed to change their diets. At the end of the eight-month study period, Dr.

Slentz found that those who exercised at the highest amount (seventeen miles) saw an 8.1 percent decrease in visceral fat.

The effect of exercise—or lack of it—was different for men than women. Of those who did not exercise, the men's weight increased 1.5 percent, while the women's weight increased 0.6 percent. However, the women's visceral fat increased by 11.6 percent—more than twice the amount of the men (5.7 percent). Those participants who exercised only a little (the equivalent of walking eleven miles per week) did not gain or lose visceral fat. According to Slentz, "The data emphasizes the high cost of continued physical inactivity, the preventive abilities of modest amounts of exercise, and the substantial benefits to be gained by performing 50 percent more exercise each week (seventeen versus eleven miles per week)."

(In general, however, exercise isn't particularly effective for weight loss, partly because we don't burn many more calories during exercise than we do simply maintaining life. For example, a 250-pound man will expend only three extra calories climbing a flight of stairs. What's more, because exercise increases your appetite, you simply eat more—something the participants in the above study weren't allowed to do. A *Nova* television show called "Marathon Challenge" dramatically showed the effect of exercise on weight. *Nova* enlisted thirteen people aged twenty-two to sixty to train for and run in the Boston Marathon. Most were sedentary types. After nine months of extensive training, all participants, with one exception, showed little change in either weight or percentage of body fat—although *Nova* didn't make a distinction between visceral and subcutaneous fat. The exception was a forty-one-year-old woman who had previously gained seventy pounds during recovery from surgery. She lost forty-five pounds—probably an amount that took her back to her presurgery weight.)

VISCERAL FAT AND STRESS

Many of us eat when we're under stress. And many of us feel stressed. At those times, we tend to go for the "comfort food"—food with lots of sugar, fat, and calories. (An A. C. Nielsen survey showed a 12 percent spike in snack food sales in the weeks following the 9/11 tragedy.) Now, scientists have discovered a link between stress and the drive to eat.

During times of stress, the brain emits a chemical signal that causes the adrenal gland to pump out large amounts of stress hormones, which are steroids such as cortisol. These hormones produce a wide array of effects designed to get your body ready for fight or flight: your heart rate, alertness, and overall body activity increases. Also, the immune system damps down. During acute stress, such as a mugging, a feedback system kicks in and shuts down the response after it's no longer needed. But during chronic stress, the cycle keeps repeating: the brain sends signals to the adrenal gland, and that gland responds by releasing stress hormones.

When the cortisol steroids are coursing through our bodies, we become active and eat high-fat and high-sugar foods. The extra calories are stored in our visceral fat, which contains special receptors that bind the cortisol steroids, thus removing them from the bloodstream. This response undoubtedly serves a useful purpose in environments of scarcity: people will run out and get food, eat it, and create stores of fat. The cortisol steroids also stimulate the visceral depots to store more fat. Some scientists believe that, because visceral fat is close by the liver, the stored energy is quickly available for conversion to energy in times of crisis. The fat, in turn, sends signals back to the brain, shutting down the production of stress hormones and making us feel better and relaxed. Unfortunately, if we then lose the fat we've accumulated, we also lose that signal to the brain that calms us down. Apparently, losing weight can thus reactivate the stress system, starting the process all over again.

HEALTH AND FAT: IT'S NOT WHAT YOU THINK

Perhaps you're better off just maintaining your normal weight. Although weight loss apparently helps diabetics, it may not be the boon to your health that you've been led to believe. According to Dr. Gary Foster, an obesity researcher at the University of Pennsylvania, "[I]f you have diabetes and you lose weight, it is likely to get better and you will go on less medication." On the other hand, many obese people do not even have insulin resistance. In a study of forty-three sedentary, postmenopausal obese women, researchers found that seventeen of them had completely normal insulin responses.

As far as other health issues are concerned, Dr. Foster isn't so sure whether losing weight will improve your health. "Will it prevent or decrease the risk of heart attack or stroke or will you be less likely to be hospitalized? We don't know that yet," he says. Dr. Rudolph Leibel, director of the division of molecular genetics at Columbia University's College of Physicians and Surgeons says that his own years of experience studying fat people had convinced him that many had none of the common risk factors that went along with cardiovascular disease—high blood sugar levels, high blood pressure, and high cholesterol levels.

Dr. Jules Hirsch, an obesity researcher at Rockefeller University, examined studies of thousands of people who lost weight. The studies kept track of who lost weight, who kept it off, who became ill, and who died. Results repeatedly showed that fat people who lost weight and kept it off had more heart disease and a higher death rate than people whose weight never changed. Dr. Katherine Flegal, an epidemiologist at the National Center for Health Statistics notes that such results "tend to indicate that weight loss is not associated with lower mortality but is actually associated with higher mortality." Such studies have raised concerns about the negative effects of weight loss and weight cycling.

So—it's hard to know what to do about that fat you're carrying around. There's no doubt that exercise is a good thing to do. I do a lot of it myself. But worrying about being overweight, a categorization discussed in the next chapter, may be counterproductive. Besides, it'll only stress you out.

Chapter 5.

WHO SAYS YOU'RE TOO FAT?

The standard device for determining whether you're underweight, average, overweight, or obese is the Body Mass Index (BMI). Even though the BMI was developed in the 1830s, it's still used—maybe overused—today for classifying people into these categories. For example the government-run China Center of Adoption Affairs requires a BMI of less than 40 for its prospective adoptive parents. So, if you're overweight and thinking about adopting a child from China, you'd better start slimming down. If you already have children and you live in Arkansas or Tennessee, you'll receive a report card noting your children's BMI numbers. Big Brother is watching your children's waistlines.

I don't set much store by the BMI device. For one thing, according to the BMI, Michael Jordan is overweight; for another, overweight people seem to live longer than thinner people.

HOW WE GOT THE BMI

The man who came up with the BMI was Adolphe Quatelet, a Belgian astronomer, mathematician, and statistician who observed, in the

1830s, that the weight of average adults was proportional to the square of their height. To confirm this observation, he measured the height and weight of French and Scottish army conscripts and found that the results yielded a standard distribution bell curve, with the thinnest at one end, the fattest at the other, and everyone else in the middle.

By performing a calculation (or using a chart or other device) using your height, weight, and a formula, you can determine your body mass (the quantity of matter in your body). The result is assigned a numerical rating, which puts you into one of the four categories—underweight, average (healthy) weight, overweight, and obese.

CALCULATING BMI USING A MATH FORMULA

To determine your BMI, divide your weight in pounds by your height in inches squared, then multiply by 703:

$$\text{BMI} = \frac{\text{weight in pounds}}{[(\text{height in inches}) \times (\text{height in inches})]} \times 703$$

For example, a person who weighs 220 pounds and is six feet three inches tall (75 inches) has a BMI of 27.5:

$$\frac{220}{(75) \times (75)} \times 703 = 27.5$$

Using this formula, a five foot five inch woman weighing 180 pounds has a BMI of 30.

CONSULTING A BMI TABLE

Instead of doing the calculations yourself, you can use the following chart (for people over age twenty). Just locate your height on the left side and your weight across the top. Your BMI is the number where the two columns intersect.

Weight (lb)

Height (ft/in)	120	130	140	150	160	170	180	190	200	210	220	230	240	250	260	270	280	290	300	310	320	330
4'5"	30	33	35	38	40	43	45	48	50	53	55	58	60	63	65	68	70	73	75	78	80	83
4'6"	29	31	34	36	39	41	43	46	48	51	53	56	58	60	63	65	68	70	72	75	77	80
4'7"	28	30	33	35	37	40	42	44	47	49	51	54	56	58	61	63	65	68	70	72	75	77
4'8"	27	29	31	34	36	38	40	43	45	47	49	52	54	56	58	61	63	65	67	70	72	74
4'9"	26	28	30	33	35	37	39	41	43	46	48	50	52	54	56	59	61	63	65	67	69	72
4'10"	25	27	29	31	34	36	38	40	42	44	46	48	50	52	54	57	59	61	63	65	67	69
4'11"	24	26	28	30	32	34	36	38	40	43	45	47	49	51	53	55	57	59	61	63	65	67
5'0"	23	25	27	29	31	33	35	37	39	41	43	45	47	49	51	53	55	57	59	61	63	65
5'1"	23	25	27	28	30	32	34	36	38	40	42	44	45	47	49	51	53	55	57	59	61	62
5'2"	22	24	26	27	29	31	33	35	37	38	40	42	44	46	48	49	51	53	55	57	59	60
5'3"	21	23	25	27	28	30	32	34	36	37	39	41	43	44	46	48	50	51	53	55	57	59
5'4"	21	22	24	26	28	29	31	33	34	36	38	40	41	43	45	46	48	50	52	53	55	57
5'5"	20	22	23	25	27	28	30	32	33	35	37	38	40	42	43	45	47	48	50	52	53	55
5'6"	19	21	23	24	26	27	29	31	32	34	36	37	39	40	42	44	45	47	49	50	52	53
5'7"	19	20	22	24	25	27	28	30	31	33	35	36	38	39	41	42	44	46	47	49	50	52
5'8"	18	20	21	23	24	26	27	29	30	32	34	35	37	38	40	41	43	44	46	47	49	50
5'9"	18	19	21	23	24	25	27	28	30	31	33	34	36	37	38	40	41	43	44	46	47	49
5'10"	17	19	20	22	23	24	26	27	29	30	32	33	35	36	37	39	40	42	43	45	46	47
5'11"	17	18	20	21	22	24	25	27	28	29	31	32	34	35	36	38	39	41	42	43	45	46
6'0"	16	18	19	20	22	23	24	26	27	29	30	31	33	34	35	37	38	39	41	42	43	45
6'1"	16	17	19	20	21	22	24	25	26	28	29	30	32	33	34	36	37	38	40	41	42	44
6'2"	15	17	18	19	21	22	23	24	26	27	28	30	31	32	33	35	36	37	39	40	41	42
6'3"	15	16	18	19	20	21	23	24	25	26	28	29	30	31	33	34	35	36	38	39	40	41
6'4"	15	16	17	18	20	21	22	23	24	26	27	28	29	30	32	33	34	35	37	38	39	40
6'5"	14	15	17	18	19	20	21	23	24	25	26	27	29	30	31	32	33	34	36	37	38	39
6'6"	14	15	16	17	19	20	21	22	23	24	25	27	28	29	30	31	32	34	35	36	37	38
6'7"	14	15	16	17	18	19	20	21	23	24	25	26	27	28	29	30	32	33	34	35	36	37
6'8"	13	14	15	17	18	19	20	21	22	23	24	25	26	28	29	30	31	32	33	34	35	36
6'9"	13	14	15	16	17	18	19	20	21	23	24	25	26	27	28	29	30	31	32	33	34	35
6'10"	13	14	15	16	17	18	19	20	21	22	23	24	25	26	27	28	29	30	31	32	34	35

Less Risk   *More Risk*

USING A WEB SITE CALCULATOR

For a more accurate and easier calculation, you can also use the automatic calculators available on Web sites such as those for the Centers for Disease Control and Prevention (www.cdc.gov) and the National Heart, Lung, and Blood Institute (www.nhlbisupport.com). You simply enter your height and weight and the computer gives the results.

INTERPRETING THE RESULTS

The World Health Organization (WHO) classifications for BMI levels are:

Less than 18.5   Underweight
18.5–24.9   Healthy
25–29   Grade 1 obesity (overweight)
30–39   Grade 2 obesity (obese)
40.0   Grade 3 obesity (morbidly obese)

According to WHO, then, the average (healthy) range for mature adults ranges from 19 to 24. For slim teenagers, an average BMI is between 18 and 22.

MAYBE THE "OVERWEIGHT" ARE HEALTHIER

The standard BMI classifications outlined above are somewhat arbitrary and even spurious. According to the WHO classifications, George W. Bush, with a BMI of 26.4, is officially overweight, along with 65 percent of Americans. Arnold Schwarzenegger, as Mr. Universe at age thirty-three, had a BMI of 33, which made him technically obese. Dr. George Bray, an obesity researcher at the Pennington Biomedical Research Center of Louisiana State University admits that President Bush's BMI doesn't make much difference to his health. Rather, he says, "Body mass is an index from which you start to make an evaluation of an individual." You need to take into consideration factors such as age, gender, physical activity, race, and where on your body the fat accumulates.

In fact, a study conducted in Norway, in the mid-1980s, which followed 1.8 million people for ten years, found that people with a BMI between 26 and 28 had the highest life expectancy. The same study found that people with a BMI between 18 and 20 (supposedly optimal), had a lower life expectancy than those with a BMI between 34 and 36. More recently, epidemiological studies in the United States report similar results. If plotted on a graph—mortality versus BMI—the graph is U-shaped, with a BMI of 25 at the bottom-most point. Looked at that way, a BMI of 25 could be seen as optimal. Another study, this one published in the August 19, 2006, issue of the *Lancet*, combined data from forty studies involving about 250,000 people with heart disease. Their results were similar to the Norway study, with overweight patients having better survival rates and fewer heart problems than those with a "normal" index number.

The March 2006 issue of *Critical Care Medicine* published a study of 1,488 patients being treated for acute lung ailments at eighty-four hospitals over a period of nearly six years. All were on ventilators. The lead researcher, Dr. James M. O'Brien Jr. of the Ohio State University Medical Center, reports "Lower BMIs [body mass indexes] were associated with higher odds of death, whereas overweight and obese BMIs were associated with lower odds." The researchers were unsure of why higher weights were associated with survival.

In November 2007, the *Journal of the American Medical Association* presented the latest data showing that overweight people have a lower death rate than people of "normal" weight. In fact, in 2004—the most recent year in which data were available—there were more than 100,000 fewer deaths among the overweight than would have been expected had they been of normal weight. According to the data, the overweight are less likely to die from a range of diseases, including Alzheimer's, Parkinson's, infections, and lung disease. What's more, the lower risk enjoyed by the overweight is not counteracted by an increased risk of dying from cancer, diabetes, or heart disease.

Clearly, the relationship between weight and death is not straightforward, except, perhaps where the overweight are concerned. According to Paul Campos, professor of law at the University of Colorado and author of *The Obesity Myth*, "The most striking data here is the remarkably consistent finding, across all [government] surveys from 1970 through 2002, that *the lowest mortality risk* is found in the 'overweight' category." Thus, if you use mortality as the criteria for determining a healthy weight, then the "overweight" category is clearly a misnomer.

WAIST-HIP MEASUREMENTS AS HEALTH PREDICTORS

While your BMI may not be a reliable indicator of heart disease risk, the way in which your fat is distributed may affect cardiovascular health. Apparently, carrying all of your excess weight in your abdomen can increase your risk for disease. It used to be that only the size of your waist mattered, and that a measurement greater than forty inches in men and thirty-five inches in women was considered dangerous. Now, the ratio of waist-to-hip measurements has been found to be a more reliable indicator of health than either body mass index or a waist-only measurement. Researchers from around the world joined to study the incidence of heart attacks (myocardial infarction) among 27,098 participants from fifty-two countries. They evaluated the relationship between heart attacks and body mass index, waist and hip circumferences, and waist-to-hip ratios. As reported in the November 5, 2005, issue of the *Lancet*, researchers concluded "waist-to-hip ratio shows a graded and highly significant association with myocardial infarction risk worldwide."

To determine your waist-to-hip ratio, measure your waist at its narrowest point and your hips at their widest point, then divide your waist measurement by your hip measurement. For example, if your waist measures 30 inches and your hips measure 36 inches, your ratio (30 ÷ 36) is 0.83. For women, the average is 0.85; for men it's 0.9. Anything above that is considered a risk for a heart attack. Researchers, who studied thousands of people, found that the waist-to-hip ratio was a predictor of heart attack even in people regarded as lean—those with a body mass index of 20, for example. The higher the waist-to-hip ratio, the higher the risk of a heart attack.

In a six-year study that tracked the health of about 15,000 people over seventy-five, during which time 6,659 died, British researchers found that those with the highest waist-to-hip ratio (closer to 1) were 40 percent more likely to die from cardiovascular disease than those with the lowest (about 0.8). In contrast, they found that a high BMI was not associated with death from circulatory disease, cancer, respiratory disease, or all causes considered together. In fact, as reported in the *American Journal of Clinical Nutrition* by Astrid E. Fletcher, the senior author of the study, "A low BMI is not desirable in older people because it may indicate risk factors such as loss of muscle or poor nutrition. Waist-to-hip ratio more accurately reflects excess body fat."

MAYBE THE OVERWEIGHT ARE HAPPIER

In a sixteen-year study that followed more than forty-five thousand male health professionals, researchers discovered that the lower the BMI, the higher the rates of suicide, and vice versa. During the time of the study, reported in the *Archives of Internal Medicine*, 131 of the men committed suicide. Compared with the men in the lowest 20 percent BMI, those in the highest one-fifth were almost 60 percent less likely to kill themselves, a "surprisingly strong relationship," according to Kenneth J. Mukamal, the lead author of the study and an associate professor of medicine at Harvard University. The authors did not have an explanation for the correlation between BMI and suicide rate of the study participants.

AN ALTERNATIVE SET OF GUIDELINES

BMI calculations don't tell the whole story of whether or not you're too fat. Total body mass depends on the density and relative mass of all body tissues, and some tissues are more dense than others. In particular, muscle is denser and takes up a smaller space than fat. A cup of muscle thus weighs more than a cup of fat. In fact, some scientists estimate that the space occupied by a pound of muscle is about 22 percent less than the space occupied by a pound of fat. Thus a five-foot-four, size ten woman with little body fat can conceivably weigh more than a size five-foot-four, size fourteen woman with lots of fat.

Considering the variations among people and their body parts, you might want to judge yourself (and Michael Jordan) on a more lenient—and perhaps more sensible—scale, such as this one developed by the USDA and US Department of Health and Human Services in 1990 (for adults over nineteen years old).

Appropriate Weights for Adults 19 and Older

Height	Total Weight Range
5'3"	107–152
5'4"	111–157
5'5"	114–162
5'6"	118–167
5'7"	121–172
5'8"	125–178
5'9"	129–183
5'10"	132–188
5'11"	136–194
6'0"	140–199

Generally, the lower weights in the chart apply to women and the higher weights to men. If your body refuses to fit within these "appropriate" ranges, and you're fit and healthy, don't worry about it. It just means you're not average.

WAYS TO MEASURE BODY FAT

Even if you maintain your weight, you may be replacing muscle with fat. In fact, Americans typically begin losing muscle tissue and gaining fat beginning about age twenty. A typical twenty-year-old man weighing 160 pounds has sixteen pounds of fat—10 percent of his body weight. But at age fifty-five, the same person at the same weight typically has forty pounds of fat—25 percent of his body weight. Fat has increased; muscle has decreased. Of course, it's unlikely that this hypothetical person would have maintained his weight over those years. On the other hand, if he were a weight lifter, he'd be more likely to maintain his twenty-year-old percentage.

If you want to find out how much of your body consists of fat and how much is other tissues, you can choose from variety of methods, all of which are reasonably accurate if properly conducted:

METHODS AVAILABLE AT
HEALTH FACILITIES AND GYMS

- *Underwater weighing*—uses Archimedes' principle to deter- mine the percentage of body fat based on your volume—that is, how much water is displaced when you're dunked under water—and your underwater weight.
- *The "Bod Pod"*—is a special chamber that measures the volume of air instead of water that your body displaces. Knowing your volume and weight, the device calculates your body density, fat, and lean mass.
- *Magnetic resonance imaging (MRI)*—is a full body scan during which you're exposed to a strong magnetic field. Hydro- gen atoms in your body respond by giving off their own mag- netic signal. Because hydrogen atoms in different tissues have slightly different signals, the MRI scanner measures those dif- ferences, which indicate body composition and which therefore

provide a picture of fat distribution, including the ratio of sub-cutaneous fat (just below the skin) to intra-abdominal fat (deeper fat).

- *DEXA*—is a whole body scan that uses two different low-dose x-rays to read bone mass and soft tissue mass. (The name stands for dual energy x-ray absorptiometry.) It can estimate the amount of lean and fat tissue and also show how fat is distrib-uted in the body.

- *Bioelectrical impedance analysis (BIA)*—is a battery-operated device that measures the resistance of your body's water to a mild electric current. Because fat acts an insulator, but lean tissue—which contains water and salt—is a good conductor, rel-ative fitness is indicated by the amount of resistance to the elec-tric current, as indicated by a voltage drop.

- *Infrared interactance*—uses a fiber optic probe with a near-infrared beam to measure the differences in light absorption of fat versus lean tissues.

- *Ultrasound*—uses high-frequency sound waves to penetrate the skin surface, pass through the fat layer, then bounce off the muscle and return to the ultrasound unit. Your body fat per-centage is determined by the amount of time it takes the ultra-sound wave to pass through the fat layer. (When I was fifty-one, I had my body fat measured using this method. My body fat per-centage was a disappointing 24.46 percent. The upside: I've never been suicidal.)

METHODS FOR HOME USE

- *Skinfold measurement*—uses special calipers to measure the skin and subcutaneous fat thickness at four specific sites on your body: front and back of the upper arm, on the back beneath the shoulder blade, and just above the hip. You (or a built-in computer) calcu-late your percentage of body fat using formulas or tables.

- *Commercial body fat analyzers*—use the same technology as the BIA analyzer described above. Some are like bathroom scales that you stand on; others are hand-held. Either way, a low electric current passes through your legs or hands. As with the BIA analyzer, body impedance technology determines the percentage of fat in your body. (I tried one of these devices at age sixty-seven and got a reading of 31 percent, seven points higher than my ultrasound measurement sixteen years earlier. Not surprising, I guess.)
- *Girth measurements*—using a tape measure to measure your waist, hips, thigh, calf, forearm, and wrist, you then apply a formula to determine percentage of body fat. (See appendix A for instructions.)

WHO'S FAT AND WHO'S THIN?

While experts purport to know how much fat is healthy and unhealthy, the truth is nobody really knows. People are different. Generally speaking, exceptionally lean males have as little as 6–10 percent body fat; for females it's 10–15 percent. Men who are considered to be fat have around 25 percent fat; for women it's 30 percent. Generally, women average in the mid- to high 20s. For women over sixty, 25 percent may be reasonable, but for a fit twenty-year-old, 17 percent body fat is more like it. The rough average fat percentage for men is 15. One technician who works at a health club has measured men who range from 2 to 53 percent—probably unhealthy percentages at either end of that spectrum. When in his racing form, Lance Armstrong's fat percentage was 4 percent. But he's a special case.

Lance Armstrong notwithstanding, you can be too thin. Our bodies need a certain amount of fat for fuel and to form important compounds such as cell membranes, and bile. This is particularly true for women, who require fat to produce fat-soluble hormones, especially estrogen. (As a rule, women should not have less than 12 percent body fat, and men 7 percent to meet these essential fat needs, but these amounts vary among individuals.) Models can even be too thin—at least those working in Madrid or Milan. In September 2006, organizers of Madrid fashion shows banned from its runways five models whose BMI was below 18.5. In Milan, fashion industry officials have also barred too-thin models from its shows. This means that supermodel Kate Moss, with a BMI of 16.4, would be out of work.

Chapter 6.

WHY YOUR BODY WANTS TO KEEP ITS SHAPE

While Kate Moss undoubtedly watches her weight, she's probably genetically programmed to be thin. Some of us are programmed to be fat. Despite mighty efforts to slim down, our weight tends to remain relatively constant. That's because our shapes are pretty much determined by our genes. It's also the reason why more than 90 percent of dieters gain back every pound they lose. Our bodies and brains send signals back and forth to regulate body weight and keep it in a narrow range. The range typically varies by only about 10 percent from a midpoint. For example, a 150-pound woman might drop to 135 pounds or rise to 165. But when her weight gets below 135 pounds, her body will begin to institute controls, including an urge to eat. When her weight rises to 165, she might develop a feeling of disgust at the sight of food, although this has never happened to me.

FIGHTING OUR GENES

Dr. Gary Foster, an obesity researcher at the University of Pennsylvania, reports that about a third of his patients regained their weight within a year of losing it; about two-thirds regained it within three years; and 80–90 percent regained it in five years. You may have had a similar experience regaining lost weight. Despite our efforts at losing weight, we tend to weigh roughly the same year after year. The fact that our weight remains relatively constant is the most obvious evidence that body fat is biologically regulated. The genetic variations among us shows up in the differences in the way people respond to diet and exercise. In studies in which everyone is given the same food and everyone exercises the same, some people lose more weight than others. Even in our environment of abundance, some people can maintain a balance between energy in (calories taken in by eating) and energy out (calories used up by exercise and maintenance of ordinary body functions). Others, however, have a chronic imbalance that favors energy in. That is, regardless of exertion, some people are genetically predisposed to be fat. You may have read about the 240-pound aerobics instructor who was refused a franchise (later reinstated) by the Jazzercise fitness company because of her weight (she is five foot eight). She's an example of someone whose body, despite her rigorous exercise schedule, refuses to relinquish its fat.

BLAME THE CAVE PEOPLE

Our genetic makeup is virtually that of our hunter-gatherer forebears. By the time they started planting crops ten thousand years ago, 100 percent of our genes were formed. In those days, the problem was finding enough to eat. When they found such a place, our early ancestors probably stuffed themselves. It was a matter of survival. Now food is always plentiful and survival is not the issue, yet we're still programmed to take advantage of high-calorie food sources. We're stuck with those pesky caveman genes.

In fact, one popular hypothesis suggests that the fattest among us probably descended from ancestors who were particularly successful at storing fat. When food was scarce, the people who hung on to their fat had a better chance of survival than their skinnier neighbors. Thus, natural selection ensured that fat people had a better chance of passing along their genes than did skinny people. As this hypothesis suggests, many of us are now stuck with genes—called "thrifty" genes—we don't need anymore. While all of us are pushing against thousands of years of evolution in our struggles to lose weight, some people are simply predisposed to store more fat than others. These unlucky people are all set for the lean times, but the lean times are unlikely to materialize.

As Ellen Ruppel Shell explains it in *The Hungry Gene*, thrifty genes are a constellation of genes—perhaps as many as two hundred—that encourage our bodies to convert calories into body fat. Because most societies have, during some periods in their history, lived under conditions of scarcity, the thrifty gene is widespread. But some populations appear to have inherited more thrifty genes than others. For example, in Kosrae, which is one of four island states that make up the Federated States of Micronesia, weather and disease at one time wiped out 90 percent of the population. Those who survived probably inherited a healthy dose of the thrifty genes—not a good thing for today's citizens of Kosrae. Now, food is both abundant and Westernized and life is easier. Inhabitants are more likely to eat SPAM than fish. Jobs in government offices have largely replaced farming and fishing. In such an environment, the once-valuable thrifty gene mechanism now predisposes citizens of Kosrae to obesity and disease. In fact, 85 percent of the population aged forty-five to sixty-four is obese. More than a third of this group suffers from high blood pressure. Ninety percent of hospital surgical admissions are diabetes related.

The Pima tribe of Native Americans in Arizona is another case in point. These people experienced crop failures and widespread starvation after white settlers, in the late nineteenth century, blocked a traditional source of irrigation water. The survivors were those with thrifty genes. While the Pima eat and exercise about the same as other Americans, about 70 percent of them are obese. Moreover, 50 percent of the tribe is now diagnosed as diabetic—the world's highest rate. These people, says Dr. Eric Ravussin, who has studied them for nearly twenty years, "suffer from a genetic disease. It's not sloth and gluttony." The Pima of Mexico, who are closely related to those of Arizona, doubtless share their thrifty genes. However, because the Pima of Mexico eat a near-starvation diet and work at labor-intensive jobs, they do not suffer from obesity, diabetes, or other associated illnesses. (However, they have a shorter life expectancy than their Arizona counterparts.)

A SECOND OPINION

According to Gary Taubes, in his book *Good Calories, Bad Calories*, the thrifty gene hypothesis doesn't hold water. For one thing, he notes, feast/famine cycles were probably not as common as we've been led to believe. For another, it's more likely that excess fat would be an evolutionary disadvantage, not an advantage. As far as feast and famines are concerned, Taubes notes that anthropological evidence as well as eyewitness testimony of early European explorers indicate that hunter/gatherer societies existed in environments in which game, insects, roots, berries, and other foods were abundant, rather than scarce. In the case of the evolutionary advantage or disadvantage of excess fat, Taubes asserts that the decreased mobility of the over-weight as well as the propensity for obesity-related diseases would definitely be disadvantageous. As he puts it, "When food is abundant, species multiply; they don't get obese and diabetic." It's more likely that such people stored enough fat to get through tough times but not enough to render them immobile.

Rather than blaming a "thrifty gene" on our propensity to hold on to fat, Taubes maintains that the real culprit is the introduction of sugar and other refined carbohydrates into our diets. In fact, the scientist who first introduced the thrifty gene hypothesis in 1962, Dr. James Neel, has since rejected his initial hypothesis and now believes that obesity in some populations is caused both by a tendency for the pancreas to oversecrete insulin and insulin resistance, both of which are triggered by, as he says, "the composition of the diet, and more specifically the use of highly refined carbohydrates." I have a hunch that it's neither just genes nor just sugar, but a combination of the two, plus other factors we have yet to discover.

INSULIN RESISTANCE

Insulin resistance—whether triggered by genes, sugar, or a combination of both—is implicated in obesity. Insulin is a hormone secreted by the pancreas that helps the body use blood sugar (glucose). Insulin does this by binding to special receptors on cells, a move that allows glucose to enter the cells. Insulin resistance occurs when the insulin receptors are blocked. Thus, glucose doesn't enter the cells as it should. Instead, it just stays in the blood. To try to overcome this resistance, the pancreas secretes additional insulin. In about a third of the people with insulin resistance, cells do not respond to even high levels of insulin and glucose continues to build up in the blood. The result is Type 2 diabetes.

To add insult to injury, insulin stimulates lipogenesis, the process that converts glucose to fat. But it inhibits lipolysis, the process that breaks down fat so it can be used for energy. Thus, insulin-resistant people are good at converting food to fat—storing excess sugar as fat instead of burning it for energy. If the thrifty gene hypothesis is true, such people have a better chance of surviving famine—and passing on their genes—than those who do not. Incidentally, many obese people don't have insulin resistance, according to Dr. Ethan Sims, an obesity expert and professor emeritus at the University of Vermont. In fact, most overweight or obese people do not have diabetes; only 5 to 10 percent have the disease, and many people with diabetes are not overweight or obese. To a large extent, Type 2 diabetes—like everything else—is genetically determined.

GENETIC PREDISPOSITION TO OBESITY

Since the discovery of the first obesity-related gene in 1994, scientists have found about fifty genes implicated in obesity, which makes it really tough if you're harboring such genes. Dr. Rudolph Leibel, an obesity researcher at Columbia Presbyterian Medical Center in New York, says, "There are powerful predispositions to being obese. The environment just makes it possible to show the genetic derangement that you've got." Thus, while the current environment certainly encourages everyone to become fatter, the fattest are those who are genetically most susceptible.

Our bodies have a complex system that regulates energy balance—both the food we take in and the burning of that food as fuel. For most obese people, an equally complex system is at work to maintain an overweight condition. Scientists have been studying this for years. For example, researchers hospitalized a 348-pound woman to study the relationship between food intake and her obesity. They calculated how many calories it would take for her to maintain that weight. For weeks, they fed her exactly that many calories, assuming her weight would stay at 348. Instead, she gained twelve pounds in two weeks. Clearly, this woman is genetically programmed to survive in a harsh environment. One of the weight-maintenance factors is metabolism: An obese person may start a dieting program with a normal metabolism, but as she begins to lose weight, her metabolism slows down. Conversely, the metabolisms of naturally thin people who purposefully try to gain weight—in the interest of science—speeds up. To gain weight, they must consume more and more calories.

Various other studies and experiments have underscored the link between genes and weight. For example, a Danish study of 540 adoptees showed that, as adults, their weight matched that of their biological parents and not their adoptive parents. Similarly, a Swedish study of identical and fraternal twins, some of whom were reared together and some apart, showed that the BMI of identical twins was nearly identical, whether they were reared together or apart. In other words, nature is more influential than nurture where fat is concerned.

Dr. Jules Hirsch, an obesity researcher at Rockefeller University, says that truly fat people really are different from people of normal weight. "There is some extraordinary genetic and environmental mix that programmed people to set for greater fat storage. Of course they overeat. But the significant issue is that they have another illness and the symptom of that is the overeating . . . a reduced fat person is not a normal person. If you take two women who both weigh 130 pounds but one used to weigh 200 pounds and one always weighed 130, they are not the same." For one thing, the person who used to weigh 200 pounds is likely to show the effects of someone who is starving: lower metabolism, constant hunger, preoccupation with food, anxiety, and depression. Even though the person is clearly not starving, she feels as if she is. Dr. Jeffrey Friedman, a researcher on the genetics of obesity at Rockefeller University writes that "the feeling of hunger is intense and, if not as potent as the drive to breathe, is probably no less powerful than the drive to drink when one is thirsty. This is the feeling the obese must resist after they have lost a significant amount of weight."

As Friedman says, "There's an illusion that you have complete control over how much you weigh—in contrast to how tall you are, or what color your eyes are, or all the other things we have to accept. The notion that there might be a biological system that evades our conscious control is not attractive to a lot of people."

HOW THOSE FAT GENES WORK

In 1994, scientists discovered a gene that is directly responsible for certain rare cases of obesity. The discovery showed that a tiny defect in a single gene can have a powerful effect on the urge to eat. This particular gene is responsible for making leptin, a hormone produced by fat cells that tells the brain that the body has had enough. If the gene is defective, people do not get the message from their brains that they can quit eating. In fact, they feel continually starved.

The first people identified as having this defect were two children, a two-year-old weighing 64 pounds and an eight-year old weighing 190 pounds. These children, who were cousins, inherited defective copies of the leptin gene, one from each parent. (Because each parent had only a single defective gene, the disease did not surface in either the mother or father.) As early as four months, the children were voraciously hungry and acted as if they were continually starving. But eating didn't quench their obsession with food. Their bodies continued to tell their brains they were starving, just as if they really were starving. The two-year-old, for example, would eat a 2,500-calorie breakfast yet still be hungry. The eight-year-old panicked whenever food was out of her sight. Putting padlocks on the cupboards and refrigerator did no good. The children scavenged through trash for bits of food.

The children's leptin level was zero. After being given injections of leptin—a therapy that must be ongoing—the children began eating normally and losing weight. Since the discovery, more people around the world have been found with the defective genes, although the condition remains a rare one. At first, doctors hoped that leptin injections could be a miracle cure for obesity. Unfortunately, such injections are effective only for those few people who lack the hormone entirely. (Leptin injections do, however, seem to help some people control their appetites.)

For those of us who do not have the defective gene—and that's nearly everyone—our fat cells release leptin into our bloodstream. As our body fat increases, so does the leptin in our bloodstream. The increased leptin tells our brain we've had enough to eat. Conversely, as our body fat decreases, less leptin is injected into our bloodstream. This drop in leptin levels tells our brain that it's time to eat.

The primary role of leptin is to keep us from getting too thin. When leptin is in short supply, our bodies respond as if starved and our appetites increase—a response that protects us when food is scarce. When we lose fat by dieting (starving), our fat cells shrink and the levels of leptin fall, a drop-off that triggers hunger pangs and increases appetite—a response that may explain the high failure rate of dieting. The mechanism that keeps weight fairly stable in most people is finely tuned, and leptin plays a major role. Generally, the more fat you have, the more leptin you make. Eduardo Nillni of Brown Univesity has discovered that a rise in leptin also sets off a sequence of hormone releases that speeds metabolism, burning calories faster. Conversely, when leptin levels drop, metabolism slows, yet another reason why it's so hard to keep weight off.

Actually, this tidy scenario of leptin-controlled appetite really only describes how animals such as squirrels regulate their appetite. With animals, as their body fat increases, leptin signals their brains to eat less and do more (perhaps hide acorns for winter). In humans, unfortunately, response to leptin signals is rarely so direct. Appetite is a complicated affair. Our appetites are sort of fail-safe systems with a complicated set of back-up strategies to ensure that we get enough to eat. Researchers have discovered more than a dozen components associated with appetite. In addition to leptin, other hormones as well as genes and receptors also play a part. For example, researchers have discovered dozens of appetite-related peptides (substances created from the breakdown of protein) in both the brain and the intestines. Another back-up mechanism is a compound called ghrelin (pronounced GREL-in). This hormone, which is produced by special cells in the stomach and in the upper part of the small intestine, sends a signal to the brain to eat whenever the stomach is empty. Ghrelin levels in the blood spike before meals and drop afterward. Thus, our urge to eat is not triggered by a single switch that is turned off and on. It appears that our appetite system is designed to make sure we get enough calories to maintain our body weight.

THE ROLE OF MICROORGANISMS

Scientists are now looking into the role that microorganisms may play in the way in which we use calories. This might strike you as implausible, but perhaps you didn't know that, of the trillions of cells in our bodies, only about one in ten belongs to us. The rest belong to the microbes—bacteria, fungi, and other tiny organisms—that inhabit all the parts of our bodies. What this means, of course, is that most of the genes we carry around don't belong to us. They belong to the microorganisms that live on and in our bodies. By the way, the combination of genes—theirs and ours—is called our "metagenome."

Most—perhaps 10 to 100 trillion—of these microorganisms live in our guts. These organisms, called microflora or microbiota, perform a wide range of services for us, including producing enzymes and aiding with digestion of food. They also help extract calories from the food we eat and help us store those calories in fat cells.

Researchers are now looking into the role these organisms may play in maintaining body fat. They have found, for example, that some common gut bacteria, *Bacteroides thetaiotaomicron* (B. theta), suppress the protein FIAF, which ordinarily prevents the body from storing fat. If FIAF is suppressed, fat accumulation increases. Researchers found that another gut microbe, *Methanobrevibacter smithii* (M. smithii), interacts with B. theta in a way that extracts additional polysaccharides from food in the gut, thus increasing the amount of calories that can be converted to fat. Dr. Jeffrey Gordon, director of the Center for Genome Sciences at Washington University, is one of the scientists studying these microbes. As a result of his studies, he has concluded, "A diet has a certain amount of absolute energy, but the amount that can be extracted from that diet may vary between individuals—not in a huge way, but if the energy balance is affected by just a few calories a day, over time that can make a big difference in body weight."

In a similar field of study, Nikhil Dhurandhar has developed a subspecialty called "infectobesity," which focuses on the role that viruses and other infectious agents may play in obesity. For example, in analyzing blood samples from fifty-two overweight patients, ten of them showed antibody evidence of prior exposure to the SMAM-1 virus, a chicken virus not previously thought to infect humans. (Interestingly, the once-infected patients had lower blood cholesterol and triglycerides than the never-infected patients.) In another study, Dhurandbhar screened 502 volunteers—360 obese and 142 not obese—for antibodies to a different virus, one called Ad-36. He found that, of those people who had been exposed to Ad-36, 30 percent of the obese group had the Ad-36 antibodies, indicating a previous exposure to the virus. The researchers don't know how a virus may affect a person's weight. Possibilities include impairment of the brain's appetite-control mechanism or inflammation that sets off a derangement in the complex system of fat regulation.

WAIT—THERE'S MORE

As if all these weight-maintaining mechanisms weren't enough, scientists keep discovering more ways in which our genes can make us fat: some people have plenty of leptin but don't process the leptin signals efficiently; some have an excess of an enzyme in muscle cells that hinders fat burning; some have a genetic defect that affects appetite-suppressing receptors or chemicals in their brains; some have a flaw in the gene responsible for converting calories into energy; and some have genes that indirectly exert some control over insulin production, a situation that can also affect fat storage.

Now it looks like how much we move is also genetically programmed. Some of us are programmed to move about less than others. In fact, the calories we burn in everyday activities are proving to be a more important factor in obesity than scientists had initially realized. Researchers at the Mayo Clinic in Rochester, Minnesota, studied twenty sedentary people—people who didn't engage in any form of exercise. Ten of these people were obese, and ten were normal weight. The group of overweight test subjects sat for about two and a half hours more per day and expended 350 fewer calories per day than the normal weight group. The study concluded that the overweight people didn't sit more because of their weight. Rather, "their movements were fixed, like it was biologically driven," according to James Levine, an endocrinologist at the Mayo Clinic. Even after losing weight, the more sedentary group maintained their sedentary habits. Regardless of their calorie intake or weight loss or gain, the test subjects continued to do the same amount of moving. In other words, when the thinner people became fatter and the fatter people became thinner, they continued to move at the same rate.

While scientists agree that metabolic rate does not predict obesity, they are finding that a factor called "nonexercise activity thermogenesis" (NEAT) is a fairly good predictor of who is fat and who is thin. NEAT is what some people call nervous energy—fidgeting types of behavior: foot tapping, finger drumming, and other types of frequent body movements. I remember a coworker who sat in a cube across from mine who was constantly jiggling his leg. He was thin. Some people actually increase these unconscious exertions in response to overeating. Dr. Levine conducted a study in which sixteen volunteers were overfed a thousand calories a day for two months. While the average weight gain was ten pounds, the range of weight gain varied dramatically among individuals. For example, one subject gained only two pounds while another gained sixteen pounds. (Not fair.) Levin concluded that the volunteers' NEAT (fidgeting activity) accounted for a tenfold difference in fat storage among the volunteers.

Then there's the stress factor. Scientists have found a neurochemical pathway that promotes fat in chronically stressed lab rats. By stimulating the pathway, researchers found that new fat deposits would accumulate, usually in the abdominal area. By blocking the pathway, they could prevent fat accumulation and shrink fat deposits. Previous studies have shown that chronic stress causes people to put on weight. Apparently, it's a survival mechanism: an impending shortage of food causes people to find food, eat, and accumulate fat. Of course, with food so plentiful, this ancestral compulsion is no longer helpful. The latest study seems to have determined the mechanism: researchers found sharply elevated levels of a substance called neuropeptide Y as well as its receptor in the fat tissue of the rats. Apparently, these substances are the culprits in stimulating fat growth of stressed animals. Of course, the pharmaceutical companies will follow this research closely.

Finally, the latest finding is that your friends can make you fat. A study published in the July 26, 2007, issue of the *New England*

Journal of Medicine tells us that if your close friend gains weight, so might you. The researchers performed a detailed analysis of a social network of 12,067 people who had been followed from 1971 to 2003 (it was actually data that had been collected as part of the large Framingham study). The researchers discovered that people were most likely to become obese when a friend became obese. I can't get too excited about this one. For one thing, such a study would be hard to replicate, and good science requires replication. For another, what are we to do with this information, shun our fat friends?

MORE STILL TO COME

Whether or not the "thrifty gene" hypothesis is right—that the fattest among us are descended from people who were most successful at storing fat—there's no denying that your genetic makeup determines your body shape. In any case, the thrifty gene hypothesis is oversimplified. The more scientists learn about our genome, the more complicated it seems to get. For example, researchers have now figured out that in addition to our genome, we've all got an *epigenome*, chemical modifiers that manage the rest of our genome by switching genes in different cells on or off. Modifications made by the epigenome are passed from generation to generation.

While the familiar double-helix with its strands of DNA is the guiding principle for making us who we are, it's these epigenome chemicals (chromatin loops, methyl groups, histones) that are responsible, during embryonic development, for differentiating one type of cell from another—making one cell develop into a heart and another into a fingernail. These chemical mechanisms are particularly sensitive to environmental signals, a capability that can alter how our genes are expressed.

For example, the chemical modifiers are probably responsible for the fact that low–birth weight babies (less than five and a half pounds) stand a good chance of becoming obese as adults. Low birth weight is usually associated with the mothers' behaviors, such as smoking, or adverse conditions, such as malnutrition. Such conclusions are the result of meticulous record keeping on various populations, such as the study of children who were born during the 1944 "Hunger Winter," when the Nazi army occupied Holland and restricted the food supply to the western part of that country. The inhabitants were forced to live on 750 calories a day or less. A typical day's worth of food might consist of a couple of slices of bread, a turnip, and one or two small potatoes.

Because birth and other records were meticulously kept, researchers were able to track 300,000 Dutch Army recruits who were in the womb during that period, tracking them from age nineteen until their sixties. The results differed depending on when malnutrition took place— throughout the pregnancy, only at the end, or only at the beginning. What they discovered is that those infants whose mothers were starved during the first trimester had higher rates of obesity by the time they reached nineteen, while the lowest rates of obesity occurred in the young men who were conceived during the end of the Hunger Winter. This study, as well as many similar studies, seems to indicate that the reduced levels of available nutrients triggers the embryo's chemical switches, setting its metabolic thermostat and/or its appetite mechanism to accommodate the nutrient shortage. The child is all set for famine but not for the feast he may find outside the womb. Other studies, especially one that followed fifty-three thousand babies born in Helsinki between 1924 and 1944, corroborates this finding. Specifically, this study showed that low–birth weight babies who rapidly gained weight after age two were the most likely to develop obesity, heart disease, and diabetes.

It seems that Mother Nature has lots of ways to push us toward pudginess, making sure we don't die of starvation. Fortunately, this is rarely an issue in our society. Unfortunately, we're better prepared for famine than we are for feast—at least in the form of Krispy Kreme donuts. Now, instead of struggling to find our next meal, we struggle with self-denial. Thanks to our differing genetic makeup, this is easier for some people than others. A degree of acceptance is called for. Despite our best efforts at shedding pounds, most of us are stuck with the shape we've got.

Chapter 7.

CHOLESTEROL CONTROVERSIES

I don't know about you, but it seems that most everyone I know is on cholesterol-lowering drugs. That's because doctors now write upward of 150 million annual prescriptions for these drugs. Actually, everyone with a cholesterol level of 240—considered by government agencies to be high—one in four Americans, including me, can be considered candidates for the drug therapy. More than ten million Americans are taking cholesterol-lowering drugs. Lipitor, the world's top-selling medicine, earns the company $13 billion a year with the help of TV ads, which, collectively, cost the pharmaceutical companies about $10 million a day.

What's more, pharmaceutical companies enlist doctors to help market their drugs—ostensibly by "educating" other doctors. In fact, one study indicates that at least 25 percent of all doctors in the United States receive drug money for lecturing at conferences, discussing drugs with other doctors, and other marketing ploys. One doctor—a psychiatrist—who described this practice wrote of receiving $750 checks for "chatting with some doctors during a lunch break." Such easy money, he continued, "left me giddy."

The numbers of people who are candidates for cholesterol-lowering drugs keep increasing. In an essay published in the *New York Times* on January 2, 2007, Drs. Gilbert Welch, Lisa Schwartz, and Steven Woloshin say that the criterion for normal cholesterol keeps dropping, such that the "disease" of high cholesterol can now be diagnosed in more than half the population. How can this many people be sick? Color me skeptical.

WHY OUR BODIES NEED CHOLESTEROL

Cholesterol is not a fat and supplies no calories to your body. Chemically, it's an alcohol, yet doesn't dissolve in water, so it's waterproof. Because it's waterproof, it makes an excellent material for building cell walls, making it possible for cells to regulate their internal environment undisturbed by what's going on around them. Waterproofing is especially critical for normal functioning of nerves and nerve cells. In fact, your brain and other parts of the nervous system contain the highest concentration of cholesterol in your body. Cholesterol constitutes over half the dry weight of your cerebral cortex. Perhaps this is why, as explained by Caroline Pond in *The Fats of Life*, people with higher blood cholesterol have been found to be capable of faster mental processing than are those with low cholesterol, and that those whose levels of blood cholesterol are unusually low, or have been artificially reduced with drugs seem to be more prone to suicide and aggressive behavior.

In addition to using cholesterol as cell wall material, your body uses it as a raw material in the healing process, especially for injured areas in the arteries and lungs. It also functions as a component of bile, which is required for the digestion of fat. What's more, cholesterol facilitates numerous biochemical processes, including mineral metabolism, serotonin uptake in the brain, and regulation of blood sugar levels. It's also used in the synthesis of various hormones, including testosterone and estrogen.

Because we can't normally eat enough cholesterol-containing

food every day to meet our bodies' needs for it, the liver and other organs manufacture cholesterol from protein, carbohydrates, and fats.

WHAT ARE LDL AND HDL?

When we eat more fat and carbohydrate than we need for energy, the liver converts these excess compounds into triglycerides, as explained in chapter 3. The triglycerides must then be transported to the various tissues in the body. To facilitate this transportation, the liver places the triglycerides into *lipoprotein* packages consisting of triglycerides, cholesterol, and protein.

Initially, the lipoprotein package has a *very low density* (VLDL), meaning that the proportion of protein in the package is very low compared with the other ingredients. (Note that lipoproteins are classified according to their densities—with the lowest-density lipoproteins having the lowest proportion of protein relative to the other package components. The less protein in a package, the more triglycerides and cholesterol.)

The liver releases the VLDLs into circulation for delivery to the various tissues—primarily muscle and fat tissue. The VLDL contains lots of triglycerides, as well as cholesterol and protein. When the packages arrive at their destination, they activate enzymes called lipoprotein lipases that release triglycerides from the VLDL packages, making the triglycerides available to the tissues for use as fuel or to be stored for later use.

Next, the VLDL is converted to *low-density lipoproteins* (LDL), which contain more cholesterol but less protein than VLDL—thus, the new density classification. The LDL packages carry triglycerides and cholesterol from cells that have an excess of these compounds toward cells and tissues that can take them up. The liver takes up most of the LDL. Some is delivered to other tissues, and about 1 percent remains in circulation to be removed by scavenger cells (macrophages). If the macrophages can remove the cholesterol fast enough, they export it into *high-density lipoprotein* (HDL) particles that carry the cholesterol back to the liver. Any cholesterol not exported via HDL may remain in the arteries and become oxidized, a condition that may lead to the development of plaque. Because of this, the cholesterol residing in LDL packages is often referred to as the "bad cholesterol." Note, however, that cholesterol is all the same, whether it's located within a VLDL, LDL, or HDL particle.

High-density lipoprotein is synthesized by both the liver and the small intestine. When newly formed, the HDLs are almost devoid of cholesterol. Thus their protein content is higher, making them denser than LDL (thus, the high-density label). However, as the HDL packages circulate about, they accumulate cholesterol either from chylomicron remnants or VLDL remnants. (Remember, chylomicrons are the lipoprotein particles created in the intestines that transport fats to the fat cells.) HDL also acquires cholesterol by extracting it from cell surface membranes, including artery walls—hence the "good cholesterol" moniker. Cholesterol-rich HDL packages then return to the liver. Once in the liver, the cholesterol in HDL may be redeployed as a component of bile salts. Because some bile remains in the feces rather than being reabsorbed, the conversion of cholesterol to bile offers a means of expelling it, thereby reducing the total amount in the body.

NOW, ABOUT THOSE CHOLESTEROL NUMBERS

In 1984, governmental agencies such as the National Heart, Lung, and Blood Institute (NHLBI) convened the "Cholesterol Consensus Conference." The conference set out to determine the cholesterol levels that should be considered unhealthy. One of the conference participants, Dr. Mary Enig, overheard three heads of the NHLBI discussing the possible outcome of these deliberations. The men were in the throes of planning a long-term trial to determine who should receive cholesterol-lowering medications. But they were having difficulty obtaining more money from Congress to conduct the trials. One of the men said to the other two: "But we can't have the cutoff at 240; it has to be at 200 or we won't have enough people to test." In other words, if the cutoff were set to 240—a number in the normal range—not enough people would qualify as having "high cholesterol." On the other hand, with the number set to 200, the researchers could find plenty of people who could meet the criteria—people whose cholesterol was above that number. The number was set at 200.

In 1985, the National Heart, Lung, and Blood Institute launched the National Cholesterol Education program. A panel of experts from this group continues to have the last word on cholesterol levels. (Incidentally, most of the members from this group have financial ties to the pharmaceutical companies that manufacture cholesterol-lowering drugs.) According to their Web site, less than 200 is desirable, 200–239 is borderline, and 240 and above is high. But according the Gary Taubes, a staff writer for *Science* magazine who had reviewed all of the information related to dietary cholesterol and listened to all the conference tapes, a total cholesterol level of 200 to 240 mg/dl is definitely in the normal range—normal meaning there was no increased risk of heart disease for men (and above which indicates a *decrease* in risk for women). You can find Taubes's paper "The Soft Science of Dietary Fat" (*Science* magazine, March 2001) on the Internet. It gives a detailed history of the deliberations and decisions that led to the guidelines.

HIGH CHOLESTEROL AS A PREDICTOR OF CARDIOVASCULAR DISEASE

Uffe Ravnskov, MD and PhD and author of *The Cholesterol Myths*, is a general practitioner in Sweden. He's also an independent researcher who has taken it upon himself to analyze the scientific literature that deals with heart disease and cholesterol. His analyses are both exacting and compelling. Among his conclusions are these:

- Half the people who have heart attacks and strokes do not have high blood cholesterol.

- People whose blood cholesterol is low develop just as many plaques in their blood vessels as people whose cholesterol is high.
- Old women with high cholesterol live longer than old women with low cholesterol.
- Cholesterol levels do not predict the risk of a heart attack in men over age sixty-five.

High total cholesterol has been shown to be a risk factor for coronary heart disease in young and middle-aged men. One explanation for this is that men of this age are in the midst of advancing their professional careers and are therefore more acutely stressed than other age groups. Mental stress (overactive adrenal glands) is a well-known cause of cardiovascular disease. (Bear in mind, however, that more than 90 percent of all cardiovascular deaths occur in people over age sixty.)

In the thirty-year follow-up of the Framingham population studies, high cholesterol was not predictive of heart disease after the age of forty-seven. In fact, the studies showed that those over fifty whose cholesterol went down had the highest risk of having a heart attack. As the authors noted: "For each 1 mg/dl drop of cholesterol there was an 11 percent *increase* [italics mine] in coronary and total mortality."

The last (and final) time I had my cholesterol checked, in 2003, it was 258. I'm not worried and I'm not taking cholesterol-lowering drugs. As I mentioned above, high cholesterol has been shown to be protective in older people.

QUESTIONING THE ROLE OF DIETARY CHOLESTEROL IN HEART DISEASE

As you're well aware, dietary cholesterol has been demonized along with saturated fat as being harmful to your heart. For years, we have been told to avoid eating food such as eggs that contain cholesterol. The truth is that there is no scientific evidence that too much dietary cholesterol promotes heart disease. If you look at the data before the

1950s, the rate of heart attacks began to increase as we began increasing our consumption of low cholesterol polyunsaturated oils and hydrogenated oils and decreasing our consumption of eggs and traditional fats such as butter and lard. For example, the consumption of margarine quadrupled between 1900 and 1950; the consumption of eggs declined by half. Dietary cholesterol has unfairly taken the rap for heart disease.

What's more, eating a low cholesterol diet may at best lower your cholesterol by a trivial 10 percent. Most people cannot change their blood cholesterol through dietary changes. (Your body produces three to four times more cholesterol than you eat and sets its own cholesterol level.) Even the Framingham study, which compared the dietary cholesterol consumed by 912 men and women with the cholesterol levels in their blood, pronounced, "There is, in short, no suggestion of any relation between diet and the subsequent development of CHD [coronary heart disease] in the study group." More than thirty studies of more than 150,000 individuals have shown that people who have had a heart attack haven't eaten more saturated fat or less polyunsaturated oil than other people.

For that matter, no research has ever shown that people who eat more eggs, each of which contain more than 200 mg of cholesterol, have more heart attacks than people who eat few eggs. Speaking of which, Dr. Ravnskov performed an interesting experiment to see the effects of egg eating on his own cholesterol. He started out the first day of his test with one egg and a cholesterol level of 278, his normal level. Each day he added another egg. After eight days, he was up to eight eggs a day (which he admits wasn't easy to do). He tested his cholesterol each day. On day eight, the day he ate eight eggs, his cholesterol was 246; down from the 278 on day one. Besides showing that egg eating didn't raise his cholesterol, he notes that "most probably, no change took place at all" because blood cholesterol measurement is not exact. If you divide a large blood sample among nine test tubes and analyze the cholesterol of each, "you will probably get nine different values," Ravnskov writes.

WHAT ABOUT THOSE CLOGGED ARTERIES?

Until recently doctors believed that heart attacks were triggered by plaques that clogged arteries. Plaques are made up of cholesterol, fatty acids, calcium, and fibrous connective tissue within the walls of arteries. While plaque is indeed embedded in damaged arterial walls, researchers no longer believe that "clogged pipes" cause heart attacks. They currently believe that cracked and ruptured artery walls trigger blood clots that block the flow of blood to the heart. In fact, the majority of what are called "clinically debilitating events" occur in areas without enough narrowing to produce angina or abnormalities you could see with a stress test.

The term scientists use to describe hardening of the arteries is "arteriosclerosis," a condition characterized by toughened areas that often contain calcium deposits (plaques). Atherosclerosis is an advanced form of arteriosclerosis. The causes of atherosclerosis are complex and not perfectly understood. Most scientists agree that the problem begins with inflammation and lesions that damage the vessel walls. Scientists don't agree on what causes the inflammation. Some believe that, for unknown reasons, white blood cells begin to attack the lining of the artery, which stimulates inflammation. In response, LDL is delivered into the arterial wall. Others believe excess cholesterol inside an artery's walls incites inflammation. Either way, what you end up with is plaque. Incidentally, autopsies have shown that arterial plaques consist of more polyunsaturated fats (41 percent) than saturated fats (26 percent).

Because atherosclerosis begins as a lesion and inflammation, the presence of C-reactive protein may be a more important indicator of coronary heart disease than high cholesterol, although this theory is now under debate. C-reactive protein is released into the bloodstream in response to inflammation and plaque buildup in the coronary arteries. People with low cholesterol can have high C-reactive protein. According to Dr. Peter Libby, chief of cardiovascular medicine at Brigham and Women's Hospital in Boston, "Inflammation is emerging as the alpha and omega of heart disease." Moreover, researchers doing autopsies have found no correlation between blood cholesterol concentration and the degree of atherosclerosis.

Some theories about the causes of damage to the lining of arteries include smoking, raised blood sugar levels, and raised levels of stress hormones. Other theories include a deficiency in fat-soluble vitamins, excess refined carbohydrates, excess omega-6 from refined vegetable oils, various vitamin and mineral deficiencies, microbes, coffee consumption, lack of exercise, exposure to carbon monoxide, and changes in the ways cause of death is reported. Some studies have shown that higher intakes of polyunsaturated fats are linked to strokes, and that older people with cholesterol below 200 are at greater risk for stroke. To further complicate matters, it looks like there's a relationship between loss of bone density and atherosclerosis. As stated by Dr. Atul Gawande in the April 30, 2007, issue of the *New Yorker*, "As we age, it's as if the calcium flows out of our skeletons and into our tissues."

In fact, scientists don't really know the cause of atherosclerosis. Even the National Heart, Lung, and Blood Institute Web page states "scientists don't know exactly how atherosclerosis begins or the exact cause." You'll just have to stay tuned—but remain skeptical.

SIZE MATTERS

Cholesterol comes in different-sized particles. It's better to have your cholesterol in large particles than small particles, which lodge themselves more easily into arterial walls than do larger particles. In fact, a predominance of the small, dense LDL (called subclass B) is associated with an increased risk of coronary heart disease. The reverse is true for the larger particles (subclass A): they're associated with a decreased risk of coronary heart disease. In fact, researchers found that heart-disease patients were three times more likely to have subclass B (small particles) than subclass A. A study in Sweden that followed 175,000 people revealed that the best predictor of heart disease risk was the dominance of the small LDL particles, which they determined by measuring the protein "cap" (apo B) that each particle wears. What's more, studies of exceptionally long-lived populations showed that their HDL and LDL particle sizes were significantly larger than "normal" people.

While genetic factors are responsible for about half the variation in LDL sizes, what you eat can affect particle size. Specifically, low-fat, high-carbohydrate diets have consistently been shown to create the smaller, denser, more dangerous particles. Of course, the reverse is true: low-carbohydrate, higher-fat diets—especially saturated fats—make for large-particle-sized cholesterol—a good thing. Note that the larger-sized particles weigh more than smaller ones, a characteristic that would increase your cholesterol numbers, since cholesterol is measured by weight, not volume (milligrams per decaliter). Thus, if your cholesterol numbers increased after a low-carbohydrate regimen, it would simply reflect the larger, heavier—but less threatening—particles.

Cholesterol particle size appears to be inversely correlated with the triglyceride levels in your blood: the bigger the cholesterol particle size, the lower the triglyceride level. While individual responses to low-carbohydrate diets vary, in general, such diets usually lower triglycerides and increase HDL in the blood (in addition to increasing cholesterol particle size). Researchers hypothesize that a low-carbohydrate diet increases the lipoprotein lipase (the enzyme that breaks down fat) produced in the muscle and fat tissue. An increase in that enzyme results in a more rapid breakdown of the fat (triglycerides), and thus a more rapid removal of the fat from circulation.

EFFECTS OF CHOLESTEROL-LOWERING DRUGS

Statin drugs such as Lipitor and Pravachol are known formally as "HMG-CoA reductase inhibitors." They work by inhibiting an enzyme, called HMG-CoA, which is required in the production of cholesterol. Cholesterol production begins with a series of steps that start with a "building block" molecule called actyl-CoA. One of the steps is the production of a chemical called mevalonate, which requires the enzyme to complete the process. Statin drugs work by inhibiting this enzyme and thus interrupting the chain of events that would normally end with cholesterol. Unfortunately, in inhibiting the production of mevalonate, statin drugs also inhibit the synthesis of other end products in the mevalonate chain, including co-enzyme Q10, a critical cellular nutrient, and dolichol, which ensures that cells respond correctly to genetically programmed instruction. Because of this, patients taking statin drugs often report serious side effects, the most common being muscle weakness.

Statin drugs have proven to lower mortality in trials for men with the highest risk of heart disease, but not by much. One trial, called CARE for "cholesterol and recurrent events," studied the use of pravastatin on four thousand people ages twenty-one to seventy-five with a past history of heart disease. For those who took a placebo, the odds of escaping death from a heart attack in five years was 94.3 percent. With the statin treatment, the odds improved to 95.4 percent. For healthy people with high cholesterol, the effect of the drugs is even smaller. A trial called WOSCOPS ("West of Scotland Coronary Prevention Study") looked at whether lowering cholesterol with statin drugs over five years would reduce fatal and nonfatal heart attacks. In this trial, the odds of escaping death were 98.4 percent for people who took the placebo and 98.8 percent for those who took the drug. Incidentally, the way in which drug companies report their statistics can be misleading. For example, in a trial where the mortality for treated people was 1.2 percent and for untreated people was 1.6 percent, the scientific papers translated the difference as a 25 percent reduction.

In the elderly as well as in women of all ages and in men without heart disease, cholesterol-lowering measures have not prevented a single death in any trial. In the cases where the drugs have proven to be protective, however, the protection occurs whether the patients' cholesterol were high or low. Interestingly, a full-page ad for Crestor, a cholesterol-lowering drug, includes the following words:

"CRESTOR is prescribed along with diet for lowering high choles-terol and has not been determined to prevent heart disease, heart attacks, or strokes."

SIDE EFFECTS OF CHOLESTEROL-LOWERING DRUGS

The University of California at San Diego has been developing a data-base of statin drug side effects. The researchers collect their data from people who report their side effects to the UCSD Web site (www .medicine.ucsd.edu/SES/contact.htm). According to the data collected so far, the best-known side effects are changes in liver function and muscle weakness, a condition called myopathy that can be very serious. A more serious and potentially fatal muscle-related side effect is called rhabdomyolysis, in which muscle fibers break down and release their contents into the bloodstream.

Other side effects reported to UCSD by statin users include prob-lems with memory, depression and irritability, headaches, joint and abdominal pain, tingling and numbness of extremities, problems with sleeping and sexual function, fatigue, dizziness, and a sense of detach-ment. There are more. Researchers are beginning to conduct tests to determine the veracity of such reports. For example, two randomized trials conducted by Dr. Matthew Muldoon at the University of Pitts-burgh showed that people on a statin drug did worse on tests of thinking and memory than people taking a placebo.

Dr. Xuemei Huang, a Parkinson's disease expert and neurologist at the University of North Carolina School of Medicine compared blood levels of LDL cholesterol in 124 Parkinson's patients with a control group of 112 of their healthy spouses. Compared with people whose LDL levels were in the upper 25 percent, those whose LDL levels were in the lower 75 percent were two and a half times as likely to suffer from Parkinson's. The men and women in the study who used cholesterol-lowering drugs were about as third more likely to have

Parkinson's as those who did not use the drugs. The researchers, who reported their findings in the December 18, 2007, issue of the *Movement Disorders* journal, recommended further tests of the effects of statins on neurodegenerative diseases.

AN EPIDEMIC OF DIAGNOSES

The high number of prescriptions for statin drugs represents what Drs. Welch, Schwartz, and Woloshin call "an epidemic of diagnoses." "While diagnoses used to be reserved for serious illness," the doctors write, "we now diagnose illness in people who have no symptoms at all, those with so-called pre-disease or those 'at risk.' " As the doctors note, an epidemic of diagnoses leads to an epidemic of treatments. Naturally, the more diagnoses, the more money for drug manufacturers, hospitals, physicians, and disease advocacy groups, not to mention the researchers and what the doctors call "the disease-based" National Institutes of Health, which secures its financing by promoting the detection of "their" diseases. "If more than half of us are sick with a diagnosis of high cholesterol," the doctors ask, "what does it mean to be normal?" Good question. No getting around it; if we aren't diseased, we are either "prediseased" or "at risk." Maybe none of us is normal.

Chapter 8.
SATURATED FATS
Healthful Food

as with cholesterol, the way saturated fat is portrayed leads you to believe that it's a kind of crud that sticks to your artery walls and gives you a heart attack. In fact, saturated fat has healthful properties that our bodies put to good use. But besides that, there's no fat in nature that consists of 100 percent saturated fat. If there were, it would be like gnawing on a candle. Similarly, totally unsaturated oils are nonexistent in natural foods. Thus, when it comes to choosing fats, you can't be a purist. You'll always get a mixture of saturated and unsaturated fats.

FAT MIXTURES: SATURATED WITH UNSATURATED

All natural fats and oils consist of a mixture of saturated, monounsaturated, and polyunsaturated fat. Even the fat in beef, which is usually referred to as "saturated," is actually less than half saturated fat. For example, the fat component of a porterhouse steak is 51 percent *monounsaturated*, of which virtually all (90 percent) is oleic acid, the same fat that's in the much-revered olive oil. Saturated fat constitutes only 45 percent of the steak's total fat, and a third of that is stearic acid—the same fat that's in chocolate. The remaining 4 percent of the fat is polyunsaturated. Likewise, lard is only 40 percent saturated fat. Olive oil, which is usually called monounsaturated, also contains polyunsaturated and saturated fat (it is 13.5 percent saturated, 73.7 percent monounsaturated, and 8.4 percent polyunsaturated).

The table below gives you an idea of the mixtures of fat types—saturated, polyunsaturated, monounsaturated—in various oils and fats.

Oil or Fat	Saturated (percentage)	Monounsaturated (percentage)	Polyunsaturated (percentage)
Beef tallow	48.4	40.5	3.1
Butter fat	61.9	28.7	3.7
Canola (rapeseed) oil	6.8	55.5	33.3
Coconut oil	86.5	5.8	1.8
Corn oil	12.7	24.2	58.0
Cottonseed oil	25.8	17.8	51.5
Flaxseed oil	9.4	20.2	66.0
Lard	39.2	45.1	11.2
Olive oil	13.2	73.6	7.9
Palm oil	48.9	37.0	9.1
Peanut oil	11.8	46.1	32.0
Safflower oil	6.8	18.6	70.1
Salmon oil	23.8	39.7	29.9
Sesame oil	14.2	45.4	40.4
Soybean oil	14.4	23.3	57.9
Sunflower oil	8.7	25.1	66.2
Walnut oil	9.1	22.8	63.3

Source: *The Omega-3 Connection,* by Andrew Stoll (Fireside, 2001)

Meats other than beef also contain less than half saturated fat. For example, the table below shows the amount of saturated fat in three ounces of various cooked meats.

Food type (3 ounces, cooked)	Saturated fat (grams)	Total fat (grams)
Beef		
Top sirloin	1.9	5.0
Flank	2.6	6.3
Ground, regular	6.2	15.5
Pork		
Ham	2.7	7.7
Spareribs	9.5	25.8
Center loin chops	2.5	6.9
Chicken		
White meat, no skin	0.9	3.0
Dark meat, no skin	2.3	8.3
Dark meat, with skin	3.0	13.4

Source: USDA National Nutrient Database for Standard Reference

The remaining fat is comprised of a combination of monounsaturated and polyunsaturated fat. For example, chicken fat is 42 percent monounsaturated and 21 percent polyunsaturated.

MIXED-UP MOLECULES

As explained in chapter 2, a triglyceride (fat) molecule consists of three fatty acids (carbon chains) connected to a glycerol backbone. The three fatty acids are nearly always different from one another, so you get some combination of saturated, monounsaturated, and polyunsaturated fatty acids in a single triglyceride molecule. For any given fat, such as lard, not every triglyceride molecule will have an identical fatty acid profile. One molecule may contain two monounsaturated fatty acids plus one saturated fatty acid; another's composition may be the reverse; while still another may contain one each of saturated, monounsaturated, and polyunsaturated fatty acid. However, even though the fatty acid content will vary from molecule to molecule, taken together the relative proportion of the various types of fatty acids remains constant. Thus, the fatty acid content of lard contains a constant proportion of approximately 40 percent saturated fatty acids,

44 percent monounsaturated fatty acids, and 10 percent polyunsaturated fatty acids.

The feature that distinguishes one type of saturated fatty acid from another is the length of its "tail"—the carbon/hydrogen chain that comprises the bulk of the molecule. The shortest of the most common saturated fatty acids, with four carbon atoms, is butyric; a long one, with eighteen carbons, is stearic. The length of the chain—designated as short, medium, or long—affects how your body digests and uses the particular fatty acids.

In general, all of the saturated fatty acids are present in milk fat—cow, goat, or human. Milk fat from cows has been shown to have more than five hundred different fatty acids and fatty acid derivatives, most of which are found only in trace amounts. Medium-chain fatty acids are prominent in tropical oils. Long-chain saturated fatty acids are present in meat fats and cocoa butter. All perform beneficial functions in our bodies. Here is a brief summary of the most common fatty acids, listed in order from short to long.

- *Butyric*—short-chain; in cow and goat milk fat.
- *Caproic*—short-chain; in cow, goat, and human milk fat and in coconut and palm kernel oils.
- *Capric*—medium-chain; in cow, goat, and human milk fat and in coconut and palm kernel oils.
- *Caprylic*—medium-chain; in coconut and palm kernel oils and in cow, goat, and human milk fat.
- *Lauric*—medium-chain; in human and cow's milk, coconut and palm kernel oils, and in genetically engineered laurate canola oil.
- *Myristic*—medium- or long-chain (the designation is debatable); in animal and vegetable fats as well as milk fats.
- *Palmitic*—long-chain; in a variety of vegetable oils (45 percent in palm oil; 14 percent in olive oil), butter fat, chicken fat, cocoa butter, lard, and human milk fat (20 to 25 percent).
- *Stearic*—long-chain; in cattle meat, dairy products, human milk, various vegetable oils, and cocoa butter (35 percent).

While a few studies have shown that particular saturated fatty acids, such as butyric or stearic, are remarkably specific in how they affect cellular metabolism, the research in this area is still rudimentary. Much remains to be learned.

WHAT ABOUT CHOLESTEROL?

To begin with, if you've read chapter 7, you should have learned that, except for a small percentage of people, diet has little effect on your cholesterol level. While scientists have spent a lot of time studying the effects of saturated fats on the lipoproteins (LDL and HDL) that carry cholesterol in the blood, few have studied the effects of individual saturated fatty acids on cholesterol. One study, for example, showed that HDL concentrations were higher after the test subjects consumed myristic acid than after they consumed stearic acid. On the other hand, other researchers found that if test subjects ate a lot of stearic acid in the form of cocoa butter, their cholesterol was reduced. It gets quite complicated, and studies keep coming up with new data.

That said, here's what researchers have learned about how each of the saturated fatty acids affects lipoproteins:

Butyric, caproic, caprylic, and capric fatty acids—are of no concern with regard to their effect on cholesterol.

Lauric and myristic fatty acid—increase both LDL and HDL concentrations.

Palmitic fatty acid—neither lowers nor raises blood cholesterol. Incidentally, when your body synthesizes fat from carbohydrates, the fat it creates is palmitic, which, in turn, can be used to synthesize other fatty acids your body needs.

Stearic fatty acids—are generally cholesterol-neutral.

In general, then, saturated fats raise HDL—a good thing. Given all the hype about eating saturated fats, you wonder what all the fuss is about.

SATURATED FAT AND HEART DISEASE

The National Institutes of Health spent several hundred million dollars trying to demonstrate a connection between eating fat and getting heart disease, but never did find the connection. The Institutes conducted five major studies, all of which revealed no such link. However, a sixth study, which was a drug trial, concluded that reducing cholesterol by drug therapy could prevent heart disease. (In this case, the probability of dying from a heart attack as a result of the drug regimen dropped from 2.0 percent to 1.6 percent.) Scientists jumped to the conclusion that a low-fat diet should reduce heart attacks. This turns out not to be the case. As the late E. H. "Pete" Ahrens, who had been a researcher on fat and cholesterol metabolism at Rockefeller University in New York City stated, "It is absolutely certain that no one can reliably predict whether a change in dietary regimens will have any effect whatsoever on the incidence of new events of [coronary heart disease] nor in whom."

To make a foolproof case that a diet high in saturated fat causes heart disease, the data must be consistent across the board: all populations eating a high saturated fat diet must have a higher rate of heart disease than populations eating a diet low in saturated fats. But this is not the case. For example, the diet of Greenland Eskimos consists of 50 percent of calories from animal fats; the diet of desert nomads consists of 10 percent of calories from animal fats. Both these groups have the lowest incidence of heart disease. Similarly, in Switzerland, death from heart attacks decreased after World War II, but during that same period, consumption of animal fat increased by 20 percent.

In thirteen studies in which the diets of patients with coronary heart disease were compared with the diets of a healthy control group, only one of the studies showed statistical significance. In that case, the patients with heart disease consumed 12.7 percent of their total calories in the form of saturated fats. In the healthy control group, the percentage of saturated fats was 12.3 percent. However, the group with heart disease (as well as similar groups in two other studies) ate more *polyunsaturated* fats than the healthy control group.

What's more, during the worst decades of the so-called "heart disease epidemic"—from the end of World War II to 1980—vegetable fat consumption increased and animal fat consumption decreased. Specifically, the vegetable-fat consumption per capita doubled from twenty-eight pounds in 1947 to fifty-five pounds in 1976, while the average consumption of all animal fat dropped from eighty-four pounds to seventy-one.

A PERSISTENT MYTH

Until recently, saturated fats and trans fats were lumped together in the various databases that researchers use to correlate dietary habits with disease. Because researchers found a correlation between disease and "saturated" fat consumption—and didn't make the distinction between saturated and trans fats—saturated fats got a bad name. While we now know better, the bad reputation lingers.

For some years now, scientists have been debunking the notion that saturated fats are a major culprit in heart disease. For example, reports in the *British Medical Journal* state that the data "do not support the strong association between intake of saturated fat and risk of coronary heart disease," and "do not support associations between intake of total fat, cholesterol, or specific types of fat and risk of stroke in men." Dr. Sylvan Lee Weinberg, past president of the American College of Cardiology writes, "Defense of the LF-HC [low-fat, high-carbohydrate] diet . . . is no longer tenable." Similarly, cardiologist Dr. L. A. Corr, and Dr. M. F. Oliver, of the National Heart and Lung Institute, London, came to the following conclusions after reviewing results of a vast number of studies: "The commonly-held belief that the best diet for the prevention of coronary heart disease is a low saturated fat, low cholesterol is not supported by the available evidence from clinical trials. In the primary prevention, such diets do not reduce the risk of myocardial infarction or coronary or all cause mortality."

After analyzing studies that focus on the "diet-heart hypothesis," researchers J. Bruce German and Cora J. Dillard state, in the *American Journal of Clinical Nutrition* (September 2004), "after 50 years of research, there was no evidence that a diet low in saturated fat prolongs life." What's more, these authors note that, "if saturated fatty acids were of no value or were harmful to humans, evolution would probably not have established within the mammary gland the means to produce saturated fatty acids—butyric, caproic, caprylic, capric, lauric, myristic, palmitic, and stearic acids—that provide a source of nourishment to ensure the growth, development, and survival of mammalian offspring."

One researcher, Dr. George Mann, called the "lipid hypothesis"—the supposed connection between saturated fat and heart disease—"the public health diversion of this century . . . the greatest scam in the history of medicine." But the myth persists.

NUTRITIONAL BENEFITS OF SATURATED FATS

Because most of the research on saturated fats has been narrowly focused on trying to find a link between consumption of saturated fatty acids and heart disease, researchers have been distracted from studying the benefits of these fatty acids. Here are a few that have come to light:

Energy source. To begin with, the long-chain saturated fatty acids (stearic and palmitic) are the preferred nutrients for the heart, as evidenced by the fact that fat around the heart muscle is highly saturated. In times of stress, your heart draws on this reserve of fat. Short- and medium-chain saturated fatty acids are a ready source of energy. Unlike long-chain fatty acids, medium-chain fatty acids (capric, caprylic, lauric), as well as the short-chain butyric and caproic fatty acids, can be absorbed directly through the intestinal wall and into the bloodstream. (As discussed in chapter 3, long-chain fatty acids must be broken down before passing through the intestinal wall, then reconstituted before delivery to cells.) If not needed as an energy source, long-chain fatty acids, which predominate in meat and dairy products, are stored in fat tissue. On the other hand, medium-chain fatty acids are quickly broken down and used primarily for energy. Thus, the fats in tropical oils seldom end up as body fat. What's more, tropical oils are also at least 2.56 percent lower in calories per gram of fat than polyunsaturated oils.

Support for chemical processes. Your cells need saturated fat to help your body perform important chemical processes and make use of vitamins and minerals. For example, your body needs saturated fat to fully convert alpha-linolenic acid (one of the essential fatty acids) into the chemicals that control body temperature, stimulate smooth muscles, and other important control functions. Also, in order for calcium to be effectively incorporated into the skeletal structure, at least 50 percent of dietary fat should be saturated. The fat in butter fat is a valuable source of A, D, K, and E (fat-soluble vitamins), vitamins that help our bodies use the minerals we eat and are essential for healthy bones and the proper development of the brain and nervous system.

Cancer prevention. Several of the short- and medium-chain fatty acids have been shown to prevent tumors. Butyric fatty acid is an interesting case in point. It's a rather unusual fatty acid, found only in ruminant (e.g., cow, goat) and human milk. No other common food fat contains this fatty acid. However, our bodies produce butyric acid in our colons through bacterial fermentation of dietary fiber. Butyric acid has been shown to play a role in cancer prevention, functioning as an antitumor agent by inhibiting abnormal cell growth and promoting natural cell death. Scientists theorize that the reason a high-fiber diet appears to be protective against colon cancer is the fact that such a diet produces butyric acid. Butyric acid also modulates the immune response and inflammation. Caprylic acid, a medium-chain fatty acid, has been reported to thwart tumors in mice.

Infection prevention. Some of the saturated fatty acids function as antibacterial and antiviral agents, especially the medium-chain fatty acids found in tropical oils, such as coconut and palm oils. The medium-chain fatty acids (and their derivatives) fight harmful bacteria, yeast, fungi, and viruses by disrupting the membranes of these organisms. Specifically, lauric acid kills a number of lipid-coated viruses and bacteria. Tropical oils are composed of about 50 percent lauric acid. For this reason, coconut oil is currently being tested as a treatment for genital herpes, hepatitis C, and HIV. Because of the health benefits of lauric acid, scientists have genetically engineered a new variety of canola oil called laurate canola that contains 36 percent lauric acid. Animal fats rarely contain lauric acid, with the exception of milk fat, which contains about 3 percent lauric acid. Unfortunately, as a part of the antisaturated fat campaign, manufacturers began (unjustifiably) replacing tropical oils with higher-calorie polyunsaturated oils, including artificially created trans fats.

Clearly, saturated fats belong in our diets. You're certainly better off eating butter than "spreads" such as I Can't Believe It's Not Butter. The ingredients in this product are listed as "liquid soybean oil and partially hydrogenated soybean oil, water, sweet cream buttermilk, salt, vegetable mono- and diglycerides, soy lecithin (potassium sorbate, sodium benzoate) as preservatives, citric acid, artificial flavor, vitamin A (palmitate), colored with beta carotene." By the way, note the "vegetable mono- and diglycerides" ingredients in this label. These are fats also, but the manufacturer doesn't let on. Instead of having three fatty acids on the glycerol backbone, a monoglyceride has one; a diglyceride has two. Nutrition labels hide the calories of these fats—which are nine per gram just like triglycerides—under the contention that "fat" consists only of triglycerides. A "nonfat" label may not be what it seems. Besides, you're better off getting a dose of anticarcinogenic caprylic fatty acid from butter than eating the "partially hydrogenated soybean oil" you'll get in I Can't Believe It's Not Butter.

Chapter 9.

OILS

Essential and Otherwise

The term "oil" usually refers to fat that's liquid at room temperature. Using this definition, oils consist primarily of unsaturated fatty acids. (Remember that all food fats contain mixtures of saturated and unsaturated fatty acids, with a predominant fatty acid that gives a fat or oil its particular characteristics.) Because the carbon chains in unsaturated fatty acids are bent, they don't stack neatly together, a characteristic that makes them liquid rather than solid.

As far as the definition is concerned, tropical oils, such as coconut and palm oils, are an exception. They're called oils, but they're composed primarily of saturated fatty acids and are liquid only when the temperature is above seventy-six degrees—temperatures common in the tropics. Had tropical oils originated in, say, Norway, where they would naturally be solid, they would have been called fats. In any case, the oils I'm talking about in this chapter consist of *un*saturated fatty acids—mono and poly.

Like saturated fats, unsaturated oils are composed of a variety of fatty acids—compounds with names such as oleic acid and linoleic acid. The various fatty acids vary according to the length of the carbon chain, the number of double bonds, and the location of the double bonds. All unsaturated fats are long or very long chain, ranging from sixteen to twenty-four carbons long.

MONOUNSATURATED OILS

As discussed in chapter 2, monounsaturated oils consist primarily of fatty acids with just a single double bond (hence, the term "mono"). These oils become firm when stored in the refrigerator. Oleic acid is the most common monounsaturated fat. Besides being the dominant fatty acid in olive oil (78 percent), oleic acid is also a major component of eggs (50 percent), butter (29 percent), and beef tallow (48 percent).

Olive oil is the best-known source of monounsaturated fat. It contains antioxidants that prevent it from becoming rancid under long-term storage. Researchers have discovered that olive oil contains a naturally occuring anti-inflammatory chemical called oleocanthal. Because inflammation is associated with heart disease, certain cancers, and other diseases, eating olive oil, with its ibuprofin-like properties, may be particularly beneficial to your health. Besides being the primary fatty acid in olive oil, oleic acid is also a major component of canola oil (65 percent), peanut oil (45 percent), and high oleic safflower oil, which comes from a plant that has undergone selective breeding to increase its oleic acid to 74 or 80 percent. Chicken, duck, goose, and turkey fat, lard, and macadamia nuts are also good sources of oleic acid.

Canola oil is also mostly monounsaturated (about 65 percent). It is made from the seed of genetically modified rape plants, which are members of the mustard family. (The name is a contraction of "Canada" and "oil," because Canada was the major producer of rapeseed oil.) The genetic modification replaces the erucic fatty acid of the original plant with oleic acid. (Erucic acid had caused undesirable changes in the hearts of test animals.) Canola oil is about 30 percent polyunsaturated fat, about 10–15 percent of which consists of omega-3 fatty acids (unless it's been refined, in which case it loses most of this ingredient).

There's some controversy about the nutritional value and safety of canola oil. While there've been no long-term studies on humans, studies on rats have been conducted since the late 1970s. Together the results indicate that canola oil is associated with fibrotic lesions in the heart. In piglets, it has been shown to cause vitamin E deficiencies and retard growth. (The FDA does not allow the use of canola oil in infant formula.) Interestingly, many studies indicate that the problems stem from high levels of omega-3 fatty acids and low levels of saturated fats. (Saturated fats, when added to the diets of test animals, mitigated the harmful effects of canola oil.) Moreover, because the high levels of omega-3 fatty acids easily become rancid and foul smelling when exposed to oxygen and high temperatures, it must be deodorized—a process that removes a large portion of the omega-3 fatty acids by turning them into trans fatty acids.

POLYUNSATURATED OILS

The important things to know about polyunsaturated oils is the fatty acids of which they are comprised:

- have two or more double bonds.
- are liquid even when refrigerated (kinks at the position of the double bonds prohibit them from packing together solidly).
- are unstable and easily become rancid.
- are divided into two families: omega-6 and omega-3.
- include the essential fatty acids required for normal cell metabolism.

The two most common polyunsaturated fatty acids in our foods are linoleic acid, with two double bonds (also called omega-6) and alpha-linolenic acid, with three double bonds (also called omega-3)—both unfortunately similarly named.

WHAT DOES "OMEGA" MEAN?

Unsaturated fatty acids are often labeled with the term "omega." When you see an "omega-3" label, for example, it's referring either to the omega-3 family of fatty acids or to a specific fatty acid in that family. Generally speaking, polyunsaturated fatty acids fall into two families: omega-6 and omega-3. (Because saturated fatty acids don't have double bonds, they aren't grouped into omega families.)

Omega, plus a number, refers to the chemical structure of the fatty acid. "Omega" refers to the last carbon at the tail end of the carbon chain, the last being the methyl end—the end opposite the glycerol backbone where the fatty acid is attached. The numbers, such as "omega-6" and "omega-3," refer to the location of the first double bond in relation to the end of the carbon chain. The diagram below shows this arrangement for two polyunsaturated fatty acids: alpha-linolenic acid, in which the first double bond occurs at the third carbon from the end, and linoleic acid, in which the first double bond occurs at the sixth carbon from the end.

Alpha-Linolenic acid fragment (omega 3)

Linoleic acid fragment (omega 6)

It's not particularly important to know these things (although they might make you feel smart). It is important, however, to be able to distinguish between the two families when you're making nutritional decisions. More about that later.

WHAT'S ESSENTIAL

Our bodies need a variety of fats to perform a variety of metabolic functions. In most cases, our bodies can synthesize a particular fat either by modifying an existing fatty acid or by creating the fatty acids from the trigycerides stored in fat cells. Our bodies cannot, however, synthesize linoleic and alpha-linolenic acids. For this reason, these two fatty acids are called *essential*: they must be supplied by the food we eat. As mentioned earlier, linoleic acid is a member of the omega-6 family; alpha-linolenic acid is a member of the omega-3 family. The table below gives a brief rundown on these two oils.

Essential oil	Sources	Minimum Daily Requirement	Comments
Linoleic acid (Omega-6)	Vegetable seed oils, such as corn, soybean, and safflower	2.3 percent of daily calories.	Are often hydrogenated, transforming them into trans fats.
Alpha-linolenic acid (Omega-3)	Flaxseed oil, fish oil, walnuts, canola oil (unhydrogenated)	0.5 to 1.5 percent of daily calories.	The parent of eicosapentaenoic (EPA) and docosahexaenoic (DHA) acids.

Four other fatty acids that are "conditionally" considered to be essential—depending on our health, diet, and genetic makeup—are gamma-linolenic acid (GLA), arachidonic acid (AA), eicosapentaenoic acid (EPA), and docosahexaenoic acid (DHA). Our bodies can make AA, DHA, and EPA from alpha-linolenic acid (ALA), but the process requires vitamin B6, magnesium, calcium, and zinc. Moreover, trans fats, cortisol, alcohol, and sugar interfere with the process. AA and DHA are essential during infancy. EPA and DHA are essential for optimal mental, visual, metabolic, and hormonal functioning.

The next table shows what percentage of the polyunsaturated portion of various fats and oils are comprised of omega-6 and omega-3 fatty acids.

Percentages of omega-6 and omega-3 fatty acids in fats

Oil or Fat	Omega-6 portion (percentage)	Omega-3 portion (percentage)		
		ALA	EPA	DHA
Beef tallow	3.1	0.6	0	0
Black currant seed oil	62.8	15.4	0	0
Borage seed oil	57.4	0	0	0
Butter fat	3.7	1.5	0	0
Canola (rapeseed) oil	33.3	11.1	0	0
Coconut oil	1.8	0	0	0
Corn oil	58.0	0.7	0	0
Cottonseed oil	51.5	0.2	0	0
Evening primrose oil	78.0	0	0	0
Flaxseed oil	66.0	53.3	0	0
Lard	11.2	1.0	0	0
Olive oil	7.9	0.6	0	0
Palm oil	9.1	0.2	0	0
Peanut oil	32.0	0	0	0
Safflower oil	70.1	0	0	0
Salmon oil	29.9	1.0	8.8	11.1
Sesame oil	40.4	0	0	0
Soybean oil	57.9	6.8	0	0
Sunflower oil	66.2	0	0	0
Walnut oil	63.3	10.4	0	0

Source: *The Omega-3 Connection*, by Andrew Stoll (Fireside, 2001)

Note that only the fish oil (in this case, salmon) contains EPA and DHA. Note also that the essential fatty acid contents of the various plant and animal fats can vary widely according to the cultivation conditions of the plants and the content of the animal feed.

ESSENTIAL OILS IN MEATS AND DAIRY PRODUCTS

Remember that meats and dairy can be important sources of both monounsaturated and polyunsaturated fatty acids. For example, palmitoleic acid, a monounsaturated fatty acid, is found almost exclusively in animal fats and has strong antimicrobial properties.

Milk fat (butter fat), which is mostly saturated fat, contains a unique *un*saturated fatty acid called conjugated linoleic acid (CLA). Only ruminant animals—those, such as cows and sheep, that digest their food multiple times—can manufacture CLA. A 1999 study of seventy-year-old males published in the *American Journal of Clinical Nutrition* showed an inverse relationship between the amount of milk fat consumed by these men and cardiovascular risk factors: "Inverse associations were found between intake of milk products and body mass index, waist circumference, LDL-HDL ratio, HDL triacylglycerols, and fasting plasma glucose. . . ." The CLA in milk fat appeared to reduce abdominal fat. Studies performed at the University of Wisconsin have also identified anticarcinogenic properties of CLA as well as properties that tend to "keep a little fat cell from getting big," as described by the researchers.

WHY ALL THE FUSS ABOUT OMEGA-3 OIL

We frequently see omega-3 oil being touted in articles and labels for its health-promoting properties. The mainstream scientific community agrees with this claim, with research to back it up. For example, articles in the *American Journal of Clinical Nutrition* and the *Journal of American Medicine* (2006) showed that people who regularly consume fish or fish oil supplements have a lower mortality and complications following heart attacks, higher HDL and lower LDL cholesterol levels, and reduced inflammation for rheumatoid arthritis sufferers.

Other studies have shown that the omega-3 oils you get from eating fish are particularly beneficial for your brain. For example, one study conducted at Harvard found that the more fish pregnant women ate during their second trimesters, the better their six-month-old infants did on tests. An Australian study reported similar findings. In this case, the study divided the test subjects (pregnant women) into two groups: one ate fish oil beginning at twenty weeks of pregnancy, and the other ate olive oil. When their children were tested at age two and a half, those of the fish-oil-eating mothers were significantly better at eye-hand coordination than those of the olive-oil-eating mothers.

Another study, published in the *Archives of Neurology*, found that elderly people who ate fish at least once a week did better on tests of memory and acuity than their peers who did not. Furthermore, they had a slower decline in mental skills each year. As reported in the *American Journal of Psychiatry* (June 2006), omega-3 supplements have been found beneficial to people with mood disorders, such as depression. Dr. Andrew Stoll, director of the Psychopharmacology Research Laboratory at McLean Hospital in Boston, has been successfully treating patients with bipolar disorder, attention deficit hyperactivity disorder (ADHD), depression, aggressive behavior, and other mental illness by prescribing fish oils, flax oil, and omega-3 supplements.

DANGERS OF POLYUNSATURATED FATS

Because the shared double bonds of unsaturated fats break more easily than the single bonds of saturated fats, unsaturated fats spoil (become rancid) more easily than the more stable saturated fats. As you can see, the more double bonds, the faster the spoilage. Thus, foods with a preponderance of polyunsaturated fats spoil the quickest, followed by foods containing mostly monounsaturated fats. In fact, the double bonds of polyunsaturated fats oxidize within a few hours of exposure to air. By contrast, foods that consist mainly of saturated fats, such as beef suet and cocoa butter, remain wholesome for months when stored at room temperature in a paper packet.

Spoilage is a result of oxidation, which occurs when an atom loses electrons, usually as a result of exposure to air, heat, or light. In fact, during processing, oils are exposed, in varying degrees, to oxygen, heat, and light, all of which cause oxidation. Atoms that have lost electrons are called free radicals. Such atoms, or groups of atoms, are highly reactive, meaning they steal electrons from other atoms to replace those they lost, a situation that produces a chain reaction of chemical changes. In stealing electrons, free radicals deplete whatever natural antioxidants (chemicals that prevent or delay oxidation) were present in the oil. Free radicals also aggressively find replacement electrons from body cell components, causing them permanent damage.

New evidence links exposure to free radicals with premature aging, with autoimmune diseases such as arthritis, and with Parkinson's, Lou Gehrig's, and Alzheimer's diseases. Free radical damage to the skin causes wrinkles and premature aging; free radical damage to the tissues and organs sets the stage for tumors; free radical damage in the blood vessels initiates the buildup of plaque. Tests and studies have repeatedly shown a high correlation between cancer and heart disease with the consumption of polyunsaturates. This may be partly due to the overconsumption of omega-6 oils relative to omega-3 oils, and partly to oxidation of the oils. By eating plenty of antioxidant fresh fruits and vegetables, you can help protect cell components, including the polyunsaturated fatty acids in cell membranes, from oxidative damage.

CHOOSING OILS

Manufacturers can get the greatest yields by using high temperatures and solvents to extract oils from plants. Such procedures expose the oils to oxidation. The best oils are those that are the least processed, which generally means extracting the oil by pressing at low temperatures. Manufacturers process oils in three ways:

- *Cold pressing:* The oil is extracted by either stone pressing or hydraulic pressing under low-intensity pressure and low (or no) heat. Cold-pressed oils contain more minerals and natural antioxidants than oils processed in other ways.

- *Expeller pressing:* The oil is extracted by squeezing the nuts, seeds, vegetables, or fruit until they release their oils, a process that usually generates enough frictional heat for it not to be considered cold pressing.
- *Solvents:* Chemical solvents are used to extract the oils from plants. Oils extracted in this way are the cheapest, but also the most oxidized, in which case some of the natural antioxidants are lost.

Obviously, then, the term to look for on bottles of oil is "cold pressing."

OLIVE OIL TERMINOLOGY

In choosing olive oil, select virgin or extra virgin made from the first pressings of the olives. These are the least oxidized. You can also check the production date on the bottles. Terms that you'll see on bottles of olive oil include:

- *Virgin:* The oil was extracted from the first cold pressing and has an acid level between 1 and 3 percent. The flavor is usually more subdued than extra-virgin oil.
- *Extra virgin:* The oil was extracted from the first cold pressing of the olives and is no more than 1 percent acid.
- *Refined:* The oil has been chemically treated to neutralize strong tastes and achieve clarity, longer shelf life and a higher smoke point (the temperature at which the oil will produce smoke when heated). Refined oil is commonly regarded as lower quality than virgin oil. A label that states *extra-virgin olive oil* or *virgin olive oil* cannot contain any refined oil.

Incidentally, where olive oil is concerned, what you see is not always what you get. Olive oil that has purportedly come from Italy may have originated in Turkey or even the United States. It may be adulterated with hazelnut oil, sunflower oil, or any number of other oils. It may have been chemically doctored to disguise unwanted flavors. It may be labeled extra-virgin, but actually be of a lesser grade. According to an article by Tom Mueller in the *New Yorker* titled "Slippery Business," "In 1997 and 1998, olive oil was the most adulterated agricultural product in the European Union." While Italy is the world's leading exporter of olive oil, Spain currently produces more olive oil than Italy, but much of it is shipped to Italy and sold as Italian oil. Because of the imported oil, local Italian farmers, who make premium oils using olives picked by hand and milled within hours, struggle to remain competitive.

HOW COOKING AFFECTS FATS

Cooking, as you would imagine, quickly degrades the polyunsaturated fatty acids in oils. Frying such oils in an open, bubbling pan both heats them enough to break the molecules into fragments and maximizes exposure to oxygen in the air. In general, the taste and digestibility of fats and oils are impaired by cooking and the nutritional quality is degraded, especially by high-temperature processes such as frying. In fact, cooking oil that is reused several times for frying may contain dangerously high concentrations of toxins. Never heat oils to the smoking point, as this not only damages their fatty acid content, but also their taste.

To preserve the nutritious properties and the flavor of unrefined oils, try the "wet sauté," a technique that shortens the time the oil is in contact with a hot pan. Start by pouring about one-fourth cup of water in the pan and heat it to just below boiling, then add the food and cook it a bit before adding the oil. Stir frequently to further reduce the time the oil is in contact with the hot metal.

As to which fats to use for which purposes, here's a guideline (from Dr. Enig):

- *Heating:* corn oil, peanut oil, olive oil. (Do not reuse the oil.)
- *Deep fat frying:* coconut oil, palm oil, lard, tallow, high oleic safflower oil, high oleic sunflower seed oil, and regular sunflower seed oil with added sesame oil and rice bran oil.
- *Salad dressings and other cold uses:* corn oil, olive oil, peanut oil.

Note that flaxseed oil and unprocessed, cold-pressed canola and soybean oils should not be heated.

STORING FATS AND OILS

Because most oils, especially unrefined oils, become rancid quickly, it's best to buy them in small quantities, and store them away from air, heat, and light in a cool dark place. It helps to buy oils in small quantities and keep them in an opaque bottle. You can store most oils in the refrigerator, although monounsaturated oils, such as olive oil, will solidify or turn cloudy. Actually, because olive oil contains antioxidants that retard spoiling, you need not store it in the refrigerator.

MAKING YOUR OWN MAYONNAISE

You can make your own mayonnaise using a blender and your choice of oil. It's easy. You probably already have the ingredients on hand: oil, eggs, vinegar, dry mustard, and salt. Here's the recipe:

1 egg
½ teaspoon dry mustard
½ teaspoon salt

2 tablespoons vinegar
1 cup vegetable oil

Place the egg in the container of an electric blender. Add mustard, salt, and vinegar. Add ¼ cup of the oil. Cover and turn the blender on to low speed. Immediately uncover and, with the motor still running, pour in the remaining oil in a thin, steady stream (do not add drop by drop). Leave the blender on about a minute after the last of the oil has been added.

You'll be doing science—making an emulsion—and you'll impress your friends and family.

Chapter 10.

WHAT'S WRONG WITH TRANS FATS

Unless you've been in a coma for the last few years, you're well aware that trans fat is the new health demon, and, with one exception, for good reason. Trans fats are oils to which hydrogen has been added in a process called hydrogenation. This artificial process gives plant-based oils the same physical properties as animal-based fats, such as butter and lard.

HOW TRANS FATS ARE MADE

To create trans fat, the manufacturer heats oil—usually cotton or sunflower oil—to about 180 degrees and pressurizes it with hydrogen in the presence of a finely powdered catalyst such as nickel. As a result, the fatty acid components of the oil become largely saturated with hydrogen atoms and the oil is transformed from liquid to solid.

$$\underset{\diagdown}{\overset{\displaystyle H \quad H}{\underset{\displaystyle C = C}{| \quad |}}} + H_2 \longrightarrow \overset{\displaystyle H \quad H}{\underset{\displaystyle H \quad H}{-C-C-}}$$

It becomes solid because the geometry of the molecule has changed to straighten out the molecules such that they can pack together to make a solid mass. Actually, the oils are only *partially* hydrogenated. Manufacturers stop the process before the oil is fully saturated with hydrogen atoms. Complete hydrogenation would make the product too much like candle wax. But in stopping the process to give the product a Crisco-like consistency, those chemical bonds that don't become hydrogenated get changed from a natural configuration to one that is not natural. And therein lies the problem. As a side effect of the hydrogenation process, a large percentage of the natural bonds on the carbon chain, called *cis* bonds, get converted to the artificial *trans* bond.

The terms cis and trans refer to the locations of the hydrogen atoms in relation to the double-bonded carbon atoms. The terms are derived from Latin stems: Cis means "on this side"; trans means "across."

Two arrangements (isomers) are possible, one natural, one artificial:

1. In the naturally occurring arrangement, both hydrogen atoms are on the *same side* of the bond:

$$\underset{\diagdown}{\overset{\displaystyle H \quad H}{\underset{\displaystyle C = C}{| \quad |}}}$$

This is called a *cis* bond. Cis bonds bend the chain of carbon atoms to form curved molecule, a feature that makes for a liquid fat (oil).

2. In the artificial arrangement, the hydrogen atoms are on *opposite* sides of the bond:

$$
\begin{array}{c}
\mathrm{H} \\
| \\
\mathrm{C}=\mathrm{C} \\
\quad | \\
\quad \mathrm{H}
\end{array}
$$

This is called a *trans* bond. Trans bonds are across from each other. They make for a straighter arrangement of carbon atoms that pack neatly side by side, a feature that makes for a solid fat.

While the differences in these two arrangements may seem trivial, changing a cis to a trans bond affects not only the fat's physical properties, such as melting temperature, but also its capacity to bind to enzymes in your body, an effect with consequences for your health.

TRANS FATS IN OUR DIETS

At the beginning of the twentieth century, Americans ate a negligible amount of trans fats. Beginning about the 1950s, believing that saturated fats were implicated in heart disease, medical organizations, such as the American Heart Association, and government agencies such as the Food and Drug Administration, urged Americans to abandon traditional fats, such as butter, in favor of partially hydrogenated oils (trans fats), such as margarine. As a result, trans fats now account for two to four percent of our daily calories (or at least they did before the current FDA guidelines went into effect). Beginning in the 1980s, the average American ate at least 12 grams of trans fats per day. Those who were avoiding animal fats typically ate more than that. Teenagers, who eat a lot of processed snack foods, were likely to eat 30 or more grams of trans fat a day. (With new awareness of the unhealthful effects of trans fats, these percentages have been steadily dropping).

Now, the USDA and FDA advise eating as little trans fat as possible. The National Academy of Science, however, concluded that there is *no* safe level of trans fat in the diet. Many scientists concur with this advice.

THE PROBLEM WITH HUMAN-MADE TRANS FATS

Trans fats compromise many bodily functions, including hormone synthesis, immune function, insulin metabolism, and tissue repair. They impair the structure and properties of cell membranes, especially those in crucial tissues such as the immune system and the brain. Because trans fatty acid molecules in the brain alter the ability of neurons to communicate, they may cause neural degeneration, diminished mental performance, and the capacity of tissues to respond effectively to injury or infection. In developing fetuses, too many trans fatty acids may impair the maturation of important nonregenerating organs such as the nervous system and the eye.

Trans fats are implicated in heart disease because our bodies respond to them by raising our LDL cholesterol and lowering our HDL cholesterol, as well as raising blood levels of triglycerides. Moreover, trans fats raise blood levels of C-reactive protein, a marker indicating a heightened inflammatory response; in other words it indicates that our body is trying to fight off infection. Trans fats can also cause hardening of the arteries under certain circumstances. One study showed that a diet deficient in magnesium combined with trans fat had the effect of increasing an influx of calcium into cells that line blood vessels, thus increasing the risk of calcification.

The *New York Times* has published several articles about trans fats, most with alarming statistics. For example, in the May 21, 2006, op-ed page, Nicholas D. Kristof states that trans fats "are estimated to kill 30,000 Americans annually and maybe more." On April 16, 2006, Nina Teiholz wrote, "a daily intake of five grams of trans fats increases the risk of contracting heart disease 4 percent to 28 percent." Speaking of alarming, the headline for the Kristof article is "Killer Girl Scouts" referring to Girl Scout cookies—which are no longer made with trans fats. The headline for the Teiholz article is "Nuggets of Death" referring to McDonald's Chicken McNuggets. Whether or not the statistics mentioned in the articles are accurate (the authors did not cite their sources), removing trans fats from our diets is the prudent thing to do.

TRANS FATS AT THE MARKET

Hydrogenated fat has been widely used for mass-produced baked goods such as crackers and cakes. In 1999, partially hydrogenated oil was present in 95 percent of the cookies, 100 percent of the crackers, and 80 percent of the frozen breakfast foods on supermarket shelves. Producers of baked goods choose hydrogenated fat because products developed with this type of fat are less susceptible to spoiling by exposure to air or sunlight than those made with unaltered fats such as butter. Hydrogenated oils also have a smoothness and high melting point that make it just the right consistency for products such as the filling in Oreo cookies. The older stick margarines are high in trans fats; some of the newer ones have none.

Now, of course, companies are scrambling to omit trans fats from their products. Crisco is a case in point. This product was introduced in 1911 as the first all-vegetable solid fat. From that time until April 2004, it was made by partially hydrogenating cottonseed oil or a mixture of soybean and cottonseed oils. Now the manufacturer, J. M. Smucker Company, is using a different method to produce what is now called Crisco Zero Grams Trans Fat Shortening. This company now creates a new product with the same properties as the old product by artificially producing a *fully* hydrogenated cottonseed oil, which is hard as nails and dry as dust, and combining it with *unhydrogenated* sunflower and soybean oils, a tricky process, I understand. Thus the product contains no partially hydrogenated oils and no trans fats. Clever.

TRANS FATS AT THE RESTAURANT

While food labels must now indicate the amount of trans fats in their products, most restaurants will do no such thing, although some cities, such as New York, are banning the use of trans fats in restaurants. The most likely place to encounter foods prepared with trans fats are fast-food chains. Because it can be reheated repeatedly, hydrogenated vegetable oil is widely used for cooking French fries. Restaurants also use it for frying meats and fish and for baking. For example, in 2000 you could expect to find trans fats in the following:

Food	Number of trans fat grams
Red Lobster Admirals Feast—one serving	22 grams
Long John Silver's Fish & More—one serving	14 grams
Burger King French Fries—one king size	8 grams
Dunkin' Donuts Old Fashioned Cake Donut—one	6 grams
Cinnabon—one	6 grams
McDonald's French Fries—large serving	4 grams
KFC Biscuit—one	4 grams
Burger King BK Big Fish Sandwich—one	4 grams
McDonald's Chicken McNuggets—9 pieces	3 grams

While restaurants, including the fast-food chains, are beginning to eliminate trans fats from their cooking, you'd do well to pay attention to the requirements in your locale.

HOW TO IDENTIFY TRANS FATS ON FOOD LABELS

Nearly ten years after the Center for Science in the Public Interest (CSPI) first asked it to do so, the Food and Drug Administration (FDA) is now requiring food manufacturers to list trans fat on nutrition facts labels, as shown in the example below:

Nutrition Facts

Serving Size 1 cup (200 g)
Servings per container 2

Amount per Serving

Calories 220   Calories from Fat 100

	% Daily Value*
Total Fat 12 g	18%
Saturated Fat 3 g	15%
Trans Fat 2 g	
Cholesterol 30 mg	10%
Sodium 235 mg	10%
Total Carbohydrate 16 g	5%
Dietary Fiber 5 g	20%
Sugars 4 g	
Protein 6 g	

Vitamin A
Vitamin C
Calcium

*Percent Daily Values are based on a 2,000
calorie diet. Your Daily Values m
or l   pending

When reading the label, consider:

- The information on each label is calculated *per serving*.
- No amount of trans fats is safe; five grams of trans fats is a lot.
- The FDA allows a product with less than half a gram per serving to be labeled as zero. Thus, if you ate two servings of an item labeled 0 trans fats, you may in fact have consumed one gram.

On food labels, you can also look for the words "partially hydrogenated" (such as partially hydrogenated soybean oil) to determine if a product contains trans fats.

NATURAL TRANS FATS

The "exception" I mentioned in the opening paragraph is the small amounts that occur naturally in the fats of ruminants—cud-chewing animals—such as cows, deer, goats, and sheep. Usually the fats of these animals (including their milk) contain about 2 percent trans fats, but can be as high as 5 percent. By contrast, partially hydrogenated vegetable oils contain as much as 50 to 60 percent of trans fat.

When cows and other ruminants eat plants, tiny microorganisms associated with the plants initially produce the enzymes for digesting the plant material. It may be that these microorganisms help maximize the nutrients obtainable from a diet of leaves, roots, and other plant materials. In the process of microbial digestion in the rumen, some unsaturated fatty acids become saturated, and some double bonds are transformed from the cis to the trans configuration. The trans double bonds that occur in natural trans fats are identical to those that occur in the synthesized trans fats. *However*, the bonds on synthetic trans fats may form on different parts of the molecule than those of naturally occurring trans fats, a situation that renders the synthetic trans fats unrecognizable to certain enzymes.

Humans are the only animals that eat significant quantities of other species' milk, and this habit arose only within the last nine thousand years when goats, sheep, and later cattle were domesticated. Thus, you could argue that eating butter, cheese, and full-cream milk—the main sources of naturally synthesized trans fatty acids in our modern diet—is no more "natural" to humans than eating artificially hydrogenated fats. That may be. But the trans fats in dairy products include a unique fatty acid called *conjugated linoleic acid* (CLA—mentioned in the last chapter), which has been found by university researchers to have anti-carcinogenic properties. Apparently the CLA ratchets down the body's production of an enzyme, called COX-2—that normally triggers inflammation. In addition to cancer, COX-2 has been indicted as a possible player in many chronic diseases, including autoimmune diseases. Moreover, naturally occurring trans fats have never been shown to share the harmful properties of synthetic trans fats. Where studies have shown the relationship between heart disease and trans fats, the trans fats have always been the human-made variety, not the naturally occurring variety. For example, the Danish Nutrition Council states that "available data suggest that ruminant trans fatty acids, especially concerning the effect on cardiovascular risk, do not possess the same unfavourable effects as industrially produced trans fatty acids. The content of trans fatty acids in industrially hydrogenated fats may reach 60 percent of the fatty acids. The equivalent number for ruminant fat is 2–5 percent."

LUMPING THE GOOD WITH THE BAD

Unfortunately all trans fats—both human-made and naturally occurring—are lumped together in the new labeling regulations. Thus, a product containing only butter will be labeled as containing trans fats. As a headline in the *New York Times* states, "Trans Fat Fight Claims Butter as a Victim." Starbucks, for example, requires all the bakers who provide its pastries to eliminate any trace of trans fats. Because no distinction is made between the two kinds of trans fats, the good is being thrown out with the bad. As a Starbucks spokesman says, "For us, it's easier for the customer to walk in and see zero grams trans fat than zero grams artificially created trans fat." According to the Food and Drug Administration rule, which went into effect in 2006, if a product has half a gram or more of trans fat per serving, the amount has to go on the label, even if butter is the only fat. So now bakers are searching for replacements, one of which is trans-fat-free margarine. As usual, people are overreacting.

AND NOW, A NEW TRANS FAT SUBSTITUTE

Because food producers are phasing out partially hydrogenated oils, now they're turning to "interesterified fats"—yet a new term we must learn. Food chemists create interesterified fats by shuffling fatty acids from one molecule to another. As you recall from chapter 2, a triglyceride molecule consists of three fatty acids attached to a glycerol backbone in an E-shaped configuration. To create interesterified fats, food chemists remove fatty acids from one triglyceride molecule and transfer them to other triglyceride molecules. Thus, they make their own blends of different kinds of fats, fats that are nonexistent in nature.

The health effects of interesterified fats have not been thoroughly investigated. Opinions differ. For example, K. C. Hayes of Brandeis University suggests that the effect of changing the positions of saturated fatty acids could potentially be on the negative side. In nature, the fatty acids being tinkered with tend to form on one end of a fat molecule or the other. But reshuffling can put them in the middle. Dr. Mary Enig says she isn't "terribly worried" about the metabolic effects of "tearing triglycerides apart and putting them back together," although she cautions that potentially hazardous chemical solvents could end up in the final product. Some preliminary studies have shown a negative effect on blood glucose and cholesterol, but these studies have been disputed.

Why not stop fooling around with fats and simply cook with butter and good old-fashioned lard? (Lard is pig fat that's been melted down and clarified.) Lard used to be the fat of choice for bakers. Unfortunately lard was demonized as an unhealthful saturated fat. As a substitute, bakers turned to Crisco, a partially hydrogenated vegetable oil, which is the newly demonized trans fat. Besides making excellent piecrust and bread, lard is a healthful fat, containing an excellent balance of 40 percent saturated and 48 percent monounsaturated fat (most of which is oleic) and 12 percent polyunsaturated, which includes linoleic fatty acids, an essential oil. (As with the fat of other animals, the amount of omega-6 and omega-3 fatty acids in lard will vary according to what has been fed to the pigs.)

Tinkering with Mother Nature often has unintended consequences. Think partially hydrogenated vegetable oil.

Chapter 11.

THE PROBLEM WITH LOW-FAT DIETS

In the early 1980s, the National Institutes of Health and other organizations decided that fat was the culprit behind the increase in heart disease and began advising all of us to restrict our fat intake. In fact, the president of the American Heart Association told *Time* magazine that if everyone went along with the low-fat idea we'd have atherosclerosis conquered by the year 2000. Thus began the anti-fat campaign, a campaign that Dr. Mann calls "the greatest biomedical error of the 20th century."

WHERE'S THE BEEF?

To kick off the campaign, the surgeon general's office declared fat the single most unwholesome component of the American diet. In 1988, that office set off to write the definitive report on the dangers of dietary fat. Assuming that the recommendations were based on sound science, the officer in charge of the report figured he'd simply collect the scientific data, have a committee of experts review the material, then publish it. The task was never completed. It turns out the committee never found the scientific data to support the idea. Or, as one of the committee members said, the subject was "too complicated." In fact, the surgeon general's office announced that the report would be killed. Award-winning science writer Gary Taubes explains all of this in "The Soft Science of Dietary Fat," published in *Science* magazine in March 2001 (you can find it on the Internet). As Taubes tells us, "Bill Harlan, a member of the oversight committee and associate director of the Office of Disease Prevention at NIH, says 'the report was initiated with a preconceived opinion of the conclusions,' but the science behind those opinions was not holding up. 'Clearly the thoughts of yesterday were not going to serve us very well.' In other words, eating a low fat diet hasn't helped us live longer."

Twenty years of low-fat recommendations have not managed to lower the incidence of heart disease in this country. As Taubes says, "Mainstream nutritional science has demonized dietary fat, yet 50 years and hundreds of millions of dollars of research have failed to prove that eating a low-fat diet will help you live longer." What's more, it looks as though, for most of us, buying into the low-fat craze hasn't made us slim either. In fact, it may have had the opposite effect.

SOME HISTORY

How did this anti-fat movement get started? In the fifties, Dr. Ancel Keys, an influential scientist and inventor of "K-rations," the prepack-aged food eaten by GIs during World War II, undertook two studies comparing the diets of Minnesota businessmen with men in seven other countries. The results of his studies implicated saturated fat and cholesterol in heart disease. His popular book, which promoted the "Mediterranean diet," was instrumental in convincing nutritionists to sideline saturated fats in their dietary recommendations. Subsequent research showed that Dr. Keys's study methods were flawed. For example, Keys had actually collected data on twenty-two countries, but because only data from seven of the countries supported his hypothesis, he tossed out the contradictory data. Despite this and other evidence disproving his notions, Keys's influence continues to this day. (At age ninety-four, Dr. Keys acknowledged that dietary cholesterol is unrelated to heart disease.)

At about the same time, the work of another researcher, David Kritchevsky, seemed to support Keys's hypothesis. Kritchevsky fed rabbits purified cholesterol, thereby inducing the growth of plaque in their arteries. Although this study was also flawed (rabbits are not built for eating fat; the plaque was different from the plaque humans get; they were fed oxidized cholesterol), manufacturers of vegetable oils took advantage of its results to convince American consumers that saturated fats should be replaced with their companies' oils. They mounted major advertising campaigns touting the health benefits of products made with their oils (e.g., "for your heart's sake"). They also worked behind the scenes to influence the dietary recommendations promoted by government agencies, such as the Food and Drug Administration. Scientists with dissenting opinions often found their research funding drying up—funding formerly supplied by such groups as the National Association of Margarine Manufacturers.

FAULTY FOOD PYRAMIDS

To motivate us to avoid fats, especially saturated fats, the Agriculture Department devised a set of nutritional guidelines, which, after much wrangling by groups such as the meat and dairy councils, was finally published as the food pyramid in 1992 (the process began in the late eighties). As you probably recall, the base of the pyramid, which indicates what foods you should eat the most, consisted of "bread, rice, cereal, pasta, 6 to 11 servings." At the top of the pyramid, indicating foods to be eaten sparingly, sits "fats, oils, sweets."

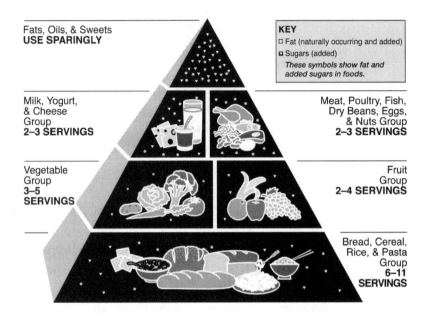

Now we know that basing our diets on bread, rice, cereal, and pasta was a bad idea. Instead of becoming healthier, we became fatter, and heart disease continued to climb along with Type 2 diabetes. We replaced meats and fats with carbohydrates. (Carbohydrates include sugars, such as table sugar and corn syrup; starchy foods such as potatoes and rice; and foods made with grains. Meats and oils do not have carbohydrates.)

In adhering to the high-carbohydrate, low-fat recommendations, we've increased our yearly consumption of grain by almost sixty pounds per person and our consumption of sweeteners, such as high-fructose corn syrup, by thirty pounds per person. We quit drinking milk and started drinking soda instead. We now drink twice as much soda as milk. Beginning in the early eighties, the incidence of obesity started rising sharply, along with Type 2 diabetes.

THE LATEST PYRAMID

Recognizing that a high-carbohydrate diet posed health problems, in 2005 the government introduced a new food pyramid (shown below). In the new pyramid, bread, rice, pasta, and potatoes no longer form the foundation. The pyramid segments are now arranged side-by-side in a vertical arrangement instead of stacked one upon the other, with grains—"at least half should be whole grain"—comprising the biggest section, and vegetables and milk forming the next-largest segments.

GRAINS   VEGETABLES   FRUITS   MILK   MEAT & BEANS

Like the first pyramid, food marketers influenced the dietary guidelines. For example, the sugar industry influenced the USDA to moderate their advice, changing it from the original "avoid too much sugar" to "choose beverages and foods to moderate your intake of sugars." In the new pyramid, fats are not labeled but are represented in the narrowest band (between "Fruits" and "Milk" in the illustration). The accompanying text, however, says that the fats we eat should be "Oils—liquid not solid." Just as we would expect: saturated fats continue to be demonized. Incidentally, until recently, saturated and trans fats were lumped together for research purposes, such that saturated fats took the rap for trans fats.

It's clear that the government should steer clear of pyramid building. The latest pyramid is heavy on the grains and allows half the grains to be refined (white bread, for example), which behave like sugar in our bodies. Grains are overrated as a dietary staple. For one thing, many people are allergic to the gluten in wheat and other grains. For another, as a relatively recent addition to the human diet (ten thousand years ago), the nutritional profile of grains, according to Dr. Loren Cordain, a Paleolithic diet specialist, is "genetically discordant" with human nutritional requirements. We've become overly dependent on grains to the exclusion of fresh vegetables and fruits. Another problem: the "oils—liquid not solid" recommendation flies in the face of current research, which shows that the excessive omega-6 fatty acids from commercial vegetable oils is deleterious to our health. In other words, in avoiding saturated fats we've started overindulging in polyunsaturated fats, as explained in chapters 8 and 9, not to mention trans fats, as explained in chapter 10.

The new pyramid also recommends drinking three glasses of *low-fat* milk per day or three servings of other dairy products per day. Low-fat milk contains oxidized cholesterol, which is not something you want to be drinking, even if you can tolerate lactose. Not only that, if you're trying to get pregnant, eating low-fat dairy products may present an obstacle to success, as discussed later in this chapter.

THE PROBLEM WITH REFINED CARBOHYDRATES

Highly refined carbohydrates, such as products made with white flour, are quickly absorbed into the blood as glucose. Within minutes after eating such food, your blood glucose spikes and you receive a surge of insulin from your pancreas. The insulin makes it possible for your cells to take up the glucose to burn for energy. A steady diet of highly refined carbohydrates may lead you to develop insulin resistance, a condition in which cells fail to respond to insulin and take up the glucose. If the cells do not take up the glucose, it remains in your blood, a condition that leads to diabetes.

Excess glucose that is not burned for energy is stored as glycogen in the liver and muscles (glycogen is stored glucose). When the liver is saturated with glycogen, the excess is converted to triglycerides, released into the bloodstream, and eventually stored as fat. Thus, you can see that eating carbohydrates—not fats—raises blood triglycerides, a well-established fact. Several studies have shown that high blood triglycerides are related to heart disease and stroke. This is partly because of the strong inverse relationship between triglyceride level and HDL cholesterol level (the higher the triglyceride level, the lower the HDL level). To lower your triglycerides, eat fewer carbohydrates.

Because insulin also regulates fat metabolism, an excess of it can make you fat. As explained in chapter 3, insulin regulates the flow of fatty acids in and out of fat cells. The in and out flow depends on whether or not your body needs it for energy, which in turn depends on the level of blood sugar available. If blood sugar is low, meaning there's no glucose available to be burned for energy, fatty acids are released to take up the slack. (This is what happens at night when you're not eating.) It's insulin that's in charge of these processes. Excess insulin shuts down the release of fat from the fat cells, thereby increasing the accumulation of fat (lipogenesis). Conversely, a deficiency of insulin releases fatty acids from the fat cells to be burned as fuel (lipolysis). Simply put, elevated insulin stimulates cells to store fat; deficient insulin stimulates cells to release fat.

An imbalance of this process is one cause of obesity. For some people, the balance is shifted toward fat storage and away from the release of fat. As Hilde Bruch, an early obesity researcher, put it, "Since it is now assumed that the genes and enzymes are closely associated, it is conceivable that people with the propensity for fat accumulation have been born with enzymes that are apt to facilitate the conversion of certain reactions in that direction."

THE IMPORTANCE OF GLYCEMIC LEVELS

Carbohydrates vary with regard to the speed with which they convert to glucose in the blood. The slower they convert to glucose, the better. (You want to avoid insulin spikes.) The term used to identify the conversion speed is "glycemic," which refers to the effect of food on the blood sugar level. The higher the glycemic level, the faster it converts. Generally speaking, the more complex the carbohydrate, the lower the glycemic index level. Thus a bagel, which almost digests itself, has a glycemic load of twenty-five (high), while oat bran, with its high fiber content, has a load of three. (Meat, which contains no carbohydrate, has a load of zero.) Orange juice has a load of fifteen; a whole orange has a load of six.

Adding fat to foods with a high glycemic index lowers the glycemic index of that food. By combining fat with carbohydrates, the food is absorbed into your bloodstream more slowly than it would be if eaten alone. So, from the standpoint of lowering the glycemic level, you're better off putting butter or cream cheese on your bagel (of course, you're also adding considerably more calories). In fact, when testing foods for glycemic index, researchers found that a Mars bar had a lower glycemic index than potatoes. Even though a Mars bar contains lots of sugar, it also contains lots of fat, which slows absorption of the sugar. It follows, then, that including fat in your meals can lower the glycemic effect of foods. (I'm not recommending replacing potatoes with Mars bars, but you might consider replacing potatoes with broccoli, for example.)

LOW-FAT MILK—BETTER THINK TWICE

The most recent government food pyramid recommends drinking three glasses of low-fat milk a day. This may not be such a good idea, especially if you're trying to get pregnant. Researchers from Harvard School of Public Health and Medical School together with Brigham and

Women's Hospital found that women who ate more than two portions a day of low-fat dairy foods were 85 percent more likely to be infertile than those who ate low-fat dairy foods less than once a week. Conversely the study showed that women who ate full-fat dairy foods, including ice cream, more than once per day had a 25 percent reduced risk of infertility compared to those who ate full-fat dairy foods only once a week.

This study followed 18,555 married, premenopausal women between the ages of twenty-four and forty-two, with no history of infertility who were either trying to become pregnant or became pregnant over an eight-year period from 1991 to 1999. The researchers concluded that "high intake of low-fat dairy foods may increase the risk of anovulatory infertility whereas intake of high-fat dairy foods may decrease this risk." Apparently low-fat dairy products cause "ovulatory disorders." The researchers, who published their findings in the *Human Reproduction* journal (February 2007) don't know why low-fat dairy and infertility may be linked. Theories include the possibility of an "unknown substance" vital for healthy ovarian functioning that requires fat for absorption. Others suggest a hormone problem created by the way milk processors convert whole milk to low-fat or skim milk.

To create the milk you buy at the supermarket, producers separate the milk as it comes from cows into its component parts (water, fat, protein, and so forth). After separation, the components are reconstituted to produce whole, low-fat, and nonfat milk. The producers use the leftover butterfat for butter, cream, cheese, and ice cream, products that earn them a higher profit than milk. To give the low-fat and skim milks body and make them more palatable, the producers add powdered milk concentrate, although this ingredient is not usually on the labels. To make the powder, they force liquid milk through a tiny hole at high pressure and temperature, then blow it out into the air. In so doing, any cholesterol in the milk is oxidized. If you recall, oxidized cholesterol is what Dr. Kritchevsky fed his plaque-ridden rabbits. Whey protein, a by-product of cheese making, is often added to low-fat dairy products to improve taste and color. In tests with mice, whey protein is suspected of producing testosterone-like effects.

LOW-FAT DIETS AND MENTAL HEALTH

Another reason to eat fat is that it makes you nicer and happier. In studying the effects of a low-fat diet on macaque monkeys, scientists discovered that such diets promoted antisocial behavior. After two years of feeding one group of monkeys a luxury diet enriched with fats and the other group a low-fat but nutritionally adequate diet, the researchers discovered that the fat-eating group groomed each other more often and spent more time within touching distance than the low-fat group. The fat-fed animals were also better able to tolerate being moved from one group to another, usually a very unsettling experience for monkeys. In other words, a high-fat, high-cholesterol diet seems to make the monkeys friendlier. In observing the antisocial behavior of the low-fat group, researchers theorized that perhaps a low-fat diet affects fats in the brain, thereby altering production of serotonin, the brain messenger molecule. A lack of serotonin is known to cause aggression.

Apparently the same is true for humans. As reported in the *British Journal of Nutrition*, January 1998, researchers tested humans in a similar way. In this case, they tested twenty healthy male and female volunteers, half of whom were put on a 41 percent fat diet and the other half were put on a 25 percent fat diet. After four weeks, the groups were swapped: the high-fat people switched to low fat and vice versa. (All meals were prepared at the university conducting the study.) At the beginning and end of each four-week period, the subjects underwent a battery of psychological tests as well as an interview by a psychiatrist who was unaware of the subjects' dietary group. The results showed that anger, hostility, and depression declined slightly during the high-fat period, but increased during the low-fat period—significantly, where anger and hostility were concerned. Researchers took pains to select volunteers they described as "psychologically robust." Had they been "feeling more stressed or were more susceptible to mental illness," according to the researchers, their "alterations in mood" may have been greater.

WEIGHT-LOSS DIETS COMPARED

Low-fat eating may also not make you thin. In 2007, Stanford University compared popular weight-loss diets, including the Zone (40 percent carbohydrates, 30 percent protein, and 30 percent fat); Ornish (high carbohydrate, very low fat); Atkins (low carbohydrate, protein and fats allowed, also low-glycemic vegetables, but fruit is restricted); and USDA's LEARN recommendations (high carbohydrate, moderately low fat). The test subjects were 311 nondiabetic, overweight women.

After one year, the women on the Atkins diet had lost the most weight. The results, which are averages for each group, are as follows: Atkins, 10.4 pounds; LEARN, 5.7 pounds; Ornish, 4.8 pounds; Zone, 3.5 pounds. More importantly the women on the Atkins diet had the largest decreases in body mass indexes, triglycerides, and blood pressure, and a higher increase in HDL cholesterol. As the report concludes, "In this study, premenopausal overweight and obese women assigned to follow the Atkins diet, which had the lowest carbohydrate intake, lost more weight and experienced more favorable overall metabolic effects at twelve months than women assigned to follow the Zone, Ornish, or LEARN diets." (Of course, after going back to their old eating habits, most of these women gained most of their weight back.)

The lead researcher, Christopher Gardner, PhD, assistant professor of medicine at the Stanford Prevention Research Center, confesses, "Many health professionals, including us, have either dismissed the value of very-low-carbohydrate diets for weight loss or been very skeptical of them, but it seems to be a viable alternative to dieters." Make no mistake, Dr. Gardner is still skeptical, wondering about the "potential long-term health problems" of Atkins-style eating. But Dr. Michael Dansinger, a Tufts-New England Medical Center researcher who conducted a similar diet comparison said, "I'm puzzled by how stubbornly nutrition authorities continue to dismiss the Atkins diet because it's counterintuitive and high in animal fat. Rather than dismissing it, we should be trying hard to learn from it."

LOW-FAT DIETS LAID TO REST—I WISH

On February 8, 2006, the front-page headline of the *New York Times* stated LOW-FAT DIET DOES NOT CUT HEALTH RISKS, REPORT SAYS. The report, which appeared in the *Journal of the American Medical Association* (JAMA), presented the long-awaited results of an eight-year study of forty-nine thousand women ages fifty to seventy-nine. Half followed a low-fat diet (20 percent of calories from fat); the other half ate all kinds of fat: butter, cream cheese, oils, and so forth. The results of the study showed that the low-fat eaters had the same rates of breast cancer, colon cancer, heart attacks, and strokes, as those who ate whatever they pleased. David A. Freeman, a statistician at the University of California, Berkeley commented, "The studies were well designed and the investigators tried to confirm popular hypotheses about the protective effect of diet against three major diseases in women. But the diet studied here turned out not to be protective after all." So, it turns out that years of low-fat eating did nothing to protect us from those diseases on which the study focused.

Even though evidence shows that people who eat low-fat diets are not happier, healthier, or thinner than people who enjoy fats, we're still admonished to avoid fats, especially the saturated variety. And, of course, low-fat foods are still highly promoted along the supermarket aisles. For me, one of the most annoying outcomes of the low-fat movement is the near disappearance of tuna packed in oil. It used to be that all tuna was packed in oil. Now most of it is packed in "spring water," which, of course, just makes it drier. To make it more palatable, you have to add more mayonnaise, which, as you know, is mostly oil.

Chapter 12.

DIVERSITY AND BALANCE

W e get plenty of advice on what to eat. But, as you've probably figured out, one diet does not fit all. Many of us (myself and most Asian people included) are lactose intolerant—we lack the enzymes that break down milk sugars. Others cannot eat anything containing gluten—an ingredient of wheat and several other grains. Gluten intolerance is called celiac disease, and, according to the Celiac Foundation Web site, it affects 1 in 133 people (it goes largely undiagnosed). Lactose intolerance and celiac disease are well-known examples of the ways individuals vary in their response to food. But it's becoming increasingly clear that the range of individual variation is both vast and complex and includes differences in the way our bodies respond to fats, cholesterol, and carbohydrates.

VARIATIONS IN DIET/GENE RESPONSES: SOME EXAMPLES

Our genes affect the ways in which we taste, metabolize, and absorb food. Conversely food can affect the way genes get expressed. All things are possible.

While most of the food-related genetic variations among us are subtle, some are more obvious and can be extremely debilitating. Remember the movie *Lorenzo's Oil*? It was about a family's search for a cure for their son Lorenzo's adrenoleukodystrophy (ALD), a degenerative disease that breaks down the myelin sheath that covers our nerves. People with ALD cannot break down very long chain fatty acids—either those in food or those made by the body. Over time, the fat molecules break down in the myelin, which itself is 80 percent fat and 20 percent protein. Interestingly, the "cure" is a special combination of two fats extracted from olive and rapeseed oil. The oil often—though not always—prevents the onset of the disease and halts its progression, although it can't repair damage that's already done.

Another rare condition, affecting about 5 percent of the US population, is familial hypercholesterolemia (FH)—a genetic disorder characterized by very high LDL cholesterol. The high cholesterol is the result of faulty receptors on the membranes of liver cells that normally remove LDL particles from blood plasma. Instead of being taken up by these cells, the excess LDL is taken up by the macrophages that constitute the earliest stage of atherosclerosis (hardening of the arteries). FH people thus develop atherosclerosis at an accelerated rate. If the person is unlucky enough to inherit two copies of the gene, he or she also develops waxy plaques in the skin, in tendons, and other locations. (FH is one condition that justifies the use of statin drugs.) While FH may be the most dramatic example, it turns out that it represents only one out of seven hundred mutations identified by scientists that affect LDL receptors.

French researchers, studying native Canadians living in an isolated community, found a subset of this population whose body mass index, percent of body fat, and fasting triglyceride level was significantly higher than others. It turns out that this group carries a gene that affects the way a protein binds with intestinal fatty acids. In other words, they metabolize fatty acids differently from others in their community. German researchers, studying a different group, found a gene that affects cholesterol concentrations in people with unusually high triglyceride concentrations. In this case, the genetic condition affects the rate at which cholesterol is eliminated through bile.

While the above examples of genetic variations may strike you as exotic, you can be sure there are plenty of less dramatic ways that food affects our genes and vice versa. For example, there's a gene that affects hepatic lipase, an enzyme that in turn affects HDL metabolism. Depending on your genetic makeup, the fat you eat will act on the enzyme such that your HDL cholesterol will either increase or decrease (happily most of us are in the HDL increasing group).

These examples—of both the rare and commonplace—demonstrate how food affects us differently. As more information emerges from studying the human genome, the more we'll learn about genetic variation. A new field of study, called nutritional genomics, tries to put genomic information to practical use. Researchers in this field study how genes and food react, with the aim of determining which foods best suit our bodies' unique requirements. Because genes are implicated in obesity, coronary artery disease, and cancers, what we eat does matter.

NUTRITIONAL GENOMICS

Food can turn some genes on and others off. At the molecular level, nutrients interact with genes by binding to certain DNA factors that regulate gene expression. Considering that a gene is essentially a recipe for making protein, the amount and type of food you eat affects the production of proteins directly. Depending on your genetic makeup, a particular food may affect your body's metabolic response differently from mine. (And don't forget about the metabolic responses of the bacteria that live in our guts.) For example, genestein, a chemical in soy, attaches to estrogen receptors and starts regulating genes. But estrogen receptors may vary from person to person, such that one person's estrogen receptors may react to genestein differently than another person's.

Ideally, if nutritional genomics were put to work on us, we would find out the best balance of food for our particular genetic makeup.

We'd then eat the right food to maintain optimum health and avoid illness. Perhaps someday this will happen. But the science of nutritional genomics is still in its infancy, and useful information is skimpy, although a few companies have begun to offer testing services. From what I've read, the number of genes they test is in the neighborhood of twenty. At the moment, what you'll probably learn from these tests is your vulnerability to insulin resistance, diabetes, bone loss, heart disease, and excess inflammation. You may also learn about your ability to metabolize certain substances, such as vitamin B_6. I understand that the advice you get tends to be fairly commonplace, such as eat more broccoli and omega-3 oils.

A BALANCE OF FATS

While we wait for the science of nutritional genomics to come of age, we must get along by listening to our bodies and by taking advantage of the best science has to offer at the moment. Where fats are concerned, it looks like our best bet is to eat saturated, monounsaturated, and polyunsaturated fats in roughly equal proportions. This is the conclusion of Dr. K. C. Hayes, professor of biology (in nutrition) at Brandeis University, who has been studying the effect of dietary fats on lipoprotein (cholesterol) metabolism for thirty-five years. His conclusions are based on an assumption that fats comprise about 30 percent of your daily calories. Intake above 40 percent, he says, raises both total and LDL cholesterol; 20 percent may lower your LDL cholesterol, but will also lower your HDL cholesterol and raise your triglyceride level, probably because an inadequate supply of essential fatty acids distorts lipoprotein metabolism.

A balance of the three categories of fats is important, Hayes says, because "neither too low saturated fatty acids nor too low polyunsaturated fatty acids is adequate, and monounsaturated fatty acids are no substitute for either. Rather one needs a balance of polyunsaturated fatty acids (to lower LDL) and saturated fatty acids (to raise HDL) for the best total cholesterol and LDL/HDL profile." Of course it becomes more complicated when you consider the different types of fatty acids, such as stearic and palmitic, as discussed in chapter 8. The best saturated fatty acid for your lipoprotein profile, Hayes says, is stearic acid (prevalent and meats and chocolate) or palmitic acid (palm oil, animal and dairy fats), although its effect on cholesterol depends on the other fatty acids it's mixed with.

I like the idea of eating saturated, monounsaturated, and polyunsaturated fats in roughly equal amounts, although it may strike you as too complicated to bother with. Part of the complication is that, as mentioned in chapter 8, all fats are mixtures of different fatty acids. You don't really know what mix you're getting in your meals. Besides, who wants to worry about all this? What's more, most of the research has focused on fats and their effects on cholesterol. Nobody really knows exactly what fats each of us requires for our own bodies' purposes. Still, balancing saturated, monounsaturated, and polyunsaturated fats strikes me as both sensible and something to shoot for (without getting overly anxious about it). The same goes for striking a balance between omega-6 and omega-3 oils.

THE IMPORTANCE OF BALANCING OMEGA-6 AND OMEGA-3 OILS

Essential fatty acids (the omega-3 family and the omega-6 family) balance the production of eicosanoids. Eicosanoids are hormonelike substances that direct a wide variety of your body's vital biochemical systems. They operate at the cellular level and are virtually invisible: they arise, do their jobs, then self-destruct. (The first eicosanoids to be discovered were isolated from the prostrate gland and were given the name prostaglandins.) The eicosanoids derived from omega-6 are different from those derived from omega-3. The two derivatives function as a sort of physiologic yin and yang. When working properly, they balance each other with opposing effects, such as:

Omega-6-derived eicosanoids	Omega-3-derived eiccosanoids
Constriction of blood vessels	Dilation of blood vessels
Higher blood pressure	Lower blood pressure
Increased blood clotting	Decreased blood clotting

As you can imagine, an imbalance in one or the other of these eicosanoids can affect your health—particularly your cardiovascular health. In order for the eicosanoids to be in balance, the omega-6 and omega-3 oils must be in balance.

We consume too little omega-3 oils compared with omega-6 oils. In the past, our ancestors got their polyunsaturated fats from the small amounts found in legumes, grains, nuts, green vegetables, fish, olive oil, and animal fats—a reasonable balance of omega-6 and omega-3 sources. At the turn of the century, most of the fatty acids our ancestors ate were either saturated or monounsaturated, primarily from butter, lard, tallows, coconut oil, and some olive oil. Now, most of the fats we eat are polyunsaturated oils derived mostly from soy, corn, and safflower oils. In fact, scientific research has shown that the amount of polyunsaturated oils we now eat constitutes about 30 percent of our calories. Two percent would be better.

THE CONSEQUENCES OF AN OMEGA-6 AND OMEGA-3 IMBALANCE

Although not all scientists agree, many believe that excess consumption of polyunsaturated oils has been shown to contribute to a large number of disease conditions including increased cancer and heart disease, immune system dysfunction; damage to the liver, reproductive organs and lung; digestive disorders; depressed learning ablity; impaired growth, and weight gain. For example, in the October 2002 issue of *Biomedicine Pharmacotherapy*, Dr. A. P. Simopoulos, president of the Center for Genetics, Nutrition and Health, asserts that "a high omega-6/omega-3 ratio, as is found in today's Western diets, promotes the pathogenesis of many diseases, including cardiovascular disease, cancer, osteoporosis, and inflammatory and autoimmune diseases, whereas increased levels of omega-3 polyunsaturated fatty acids (PUFA) (a lower omega-6/omega-3 ratio), exert suppressive effects."

WHY THE IMBALANCE

We simply eat more foods containing omega-6 than omega-3 oils. For one thing, omega-6 oils are more commonplace in our diets. For example, soybean oil is the most commercially abundant food fat in the United States and represents nearly 80 percent of the seed oils used in food processing. It's commonly used for mayonnaise, among other processed foods. Three-quarters of the soybean oil produced has been partially hydrogenated, meaning it has been turned into a trans fat, a process that diminishes any omega-3 oils it had.

On the other hand, we're less likely to eat as much of the foods in which omega-3 oils are abundant, such as flaxseed oil, salmon oil, and walnut oil. For this reason, it requires a little more diligence to make sure you get enough omega-3 fatty acids, especially the DHA variety found only in fish oil. As a rule, the leaves of plants contain more of the omega-3 fatty acids, and the seeds contain more omega-6. We eat far more foods produced from seeds, such as flour and oils, than leaves.

Modern food production has helped to tip the balance away from omega-3 and toward omega-6. For example, because omega-3 fatty acids spoil more readily than omega-6, growers have selected for plants that producer fewer omega-3 fatty acids. What's more, when we partly hydrogenate oils, we destroy their omega-3 components. It doesn't help that official dietary advice since the 1970s has promoted the consumption of oils high in omega-6, such as corn and soy.

At the most, the ratio should be three or five omega-6 fatty acids to one of omega-3. Many in the scientific community recommend a one to one ratio. We're way off on that score, with a current consumption ratio of anywhere from 10:1 to 30:1 omega-6 to omega-3. We need a balance of the two. The only way to ensure a proper balance of the essential oils is to consume both oils—omega-6 and omega-3—in approximately equal amounts.

STRIKING THE ESSENTIAL OIL BALANCE

Dr. Mary G. Enig, author of *Know Your Fats*, offers the following suggestions for amounts of omega-3 and omega-6 oils to eat on a daily basis, based on a two thousand–calorie diet. To get 2.2 to 3.3 grams of *omega-3 alpha linolenic acid*, you can choose among the following: 1.5 teaspoons flaxseed oil, 1.5 tablespoons canola oil, 2 tablespoons walnut oil, or 3.0 tablespoons soybean oil or whole soybeans. You can also obtain omega-3 fatty acid by eating flaxseeds, walnuts, or soybeans in about the same amounts as the oil.

To get the EPA (eicosapentaenoic) and DHA (docosahexaenoic) omega-3 fatty acids found mostly in fish, some authorities suggest eating two to three grams of fish oil, either in the form of capsules or fish. For each 3.5-ounce serving, you can expect to be eating the following amounts of oil:

Cod, flounder, haddock, and halibut: .5 grams
Anchovy, mullet, salmon: 1 to 1.5 grams
Herring, lake trout: 1.5 to 2 grams
Mackerel, sardines: 2 to 3.3 grams

To get 4.4 to 6.7 grams per day *omega-6* (*linoleic acid*), you can choose among the following: 2 teaspoons sunflower oil, 1 tablespoon corn oil, 2.0 tablespoons borage oil, 2 tablespoons flaxseed oil, 2.5 tablespoons canola oil, or 5 tablespoons olive oil.

Admittedly, most of us would be hard pressed to follow such a regimen. Still, it's good to have these suggestions to guide us.

OUR BALANCING MECHANISMS

Our bodies have regulating mechanisms to keep us in balance. We sweat when we're hot and shiver when we're cold. Hunger impels us to eat; thirst moves us to drink. Most of these balancing activities, however, take place outside of our consciousness—activities such as the movement of molecules in and out of our cells. Fat cells are a case in point. As with other body systems, our fat cells maintain equilibrium between the forces that deposit fat and the forces that release fat. As discussed in chapter 3, when we've eaten more food than we need for energy, fat is deposited in the cells; when we've gone for a period without food, fat is released. The filling and emptying is regulated by feedback systems, chiefly the nervous system residing in the hypothalamus part of the brain and the endocrine (hormone) system. Among the hormones, insulin is the chief regulator that determines whether fat is deposited or released, as explained in chapter 11.

Sometimes our balancing systems are out of whack. For example, there's a rare condition called diabetes insipidus in which the kidneys don't get the hormonal message to hang on to water. People who have this disease are constantly thirsty. They drink gallons of water but also excrete gallons of water, so they're actually dehydrated. Similarly—but perhaps not so dramatically—our fat balancing systems can also be out of whack. Obese people tend to have a hormonal imbalance that predisposes them to accumulate excess fat. That is, their disorder is one of fat accumulation, not one of overeating. Overeating is a side effect of an underlying metabolic defect in which the fat tissue doesn't release the calories fast enough to satisfy the needs of the cells in their other tissues between meals. For this reason, obese people are stimulated to eat extra calories to maintain what for them is a balanced system. In other words, their cells can be in a semistarved condition, even if the person is clearly not starving.

When you think about it, you can see that all of us—fat or thin— have a balancing system that tends to keep us at a rather consistent weight for many years. Over the course of twenty years, for example, you may consume nearly twenty million calories—more than twenty-five tons of food. It's unlikely that every day you carefully monitor the calories you eat to balance the energy you expend such that your weight remains relatively stable, and yet it does. Your body sees to it. Unfortunately, as Gary Taubes writes in *Good Calories, Bad Calories*, "neither eating less nor exercising more will lead to long-term weight loss, as the body naturally compensates. We get hungry, and if we can't satisfy that hunger, we'll get lethargic and our metabolism will slow down to balance our intake. This happens whether we're lean or obese."

Generally speaking, lean people are more active than fat people because a greater proportion of the food they consume is made available to their cells and tissues for energy. With energy to burn, they're more inclined to be restless and impelled to be physically active. The opposite is true of obese people: the calories they consume go to making fat rather than to burn for energy. On a calorie-restricted diet, both fat and thin people become hungry and lethargic for the same reason: their cells are not receiving enough nutrients to maintain their natural balance.

An interesting example of the way we differ in our response to food as well as the way our bodies try to maintain a stable weight is an experiment performed on convicts in 1960. The researcher, Ethan Sims at the University of Vermont, overfed eight convicts, starting them at 4,000 calories a day, then increasing their food allotment until they were eating 10,000 calories a day. Of his eight test subjects, all of whom were sedentary, two gained weight easily and six did not. One gained less than ten pounds after thirty weeks of high-calorie eating (he went from 124 pounds to 143). When the experiment ended all the men lost the weight they gained, just as obese patients typically return to their usual weights after being on semistarvation diets. Whether we eat too much or too little, our bodies try to keep us in balance.

Chapter 13.

FAT, THE FARM, AND FAMILY

everyone says we're too fat. We not only worry about our own weight, we worry about everybody else's. Apparently an obesity epidemic is afoot. The experts blame fast food, overly large portion sizes, too much fat and sugar in our diet, too few fruits and vegetables in our diet, and too little exercise. While this may all be true (except, perhaps for the fat part), I'd bring a slightly different perspective to the discussion, one that has more to do with our society and its institutions than with individuals and their lack of restraint.

RIVERS OF CORN SYRUP

An article titled "The Pediatric Obesity Epidemic: Causes and Controversies" by Arnold H. Slyper in the *Journal of Clinical Endocrinology and Metabolism* concludes that the increase in consumption of carbohydrates is the real cause of our "obesity epidemic." Slyper notes that, while "fat consumption in American children has fallen over the last 3 decades," the "amount of carbohydrate in children's diets has been increasing." Specifically, in the years between 1978 and

1994, teenage boys have tripled their consumption of soft drinks. Indeed, David Ludwig, director of the obesity program at Children's Hospital in Boston, reports that children arriving at his clinic consume as many as a thousand calories a day in soft drinks. The mother of one of these boys, who drinks a couple of big bottles of soda with his dinner notes that this is "what any boy his age would drink, you can't expect him to settle for water."

Soft drinks now contribute about 8 percent of calories consumed by adolescents. The calorie culprit, in this case, is high-fructose corn syrup (HFCS). HFCS was not available until 1980. Since then, it has become the leading source of sweetness in our diet. According to Michael Pollan in his book *The Omnivore's Dilemma*, Americans' annual consumption of HFCS has gone from forty-five pounds in 1985 to sixty-six pounds today. Unfortunately, HFCS didn't simply replace other sugars, such as cane. It was added. Now, our consumption of all added sugars stands at 158 pounds per person, a thirty-pound increase since 1985. Of the sixty-six pounds of HFCS that we consume each year, most of it is in the form of soft drinks.

MOUNTAINS OF CORN

Every year food processors convert 530 million bushels of corn into 17.5 billion pounds of high-fructose corn syrup. That's a lot of corn. The overproduction of corn was triggered by agricultural policies enacted in the seventies that were designed to drive up yields and drive down the price of certain agricultural commodities, especially corn and soybeans. So now we have mountains of cheap corn, which accounts for most of the surplus calories we're eating. Cheap corn makes for cheap sweeteners. In 1980, corn first became an ingredient in Coca-Cola; by 1984 Coca-Cola and Pepsi had switched over entirely from sugar to HFCS. By using HFCS as the sweetener, soft drink manufacturers could save money on every bottle they produced. But instead of cutting the price of a bottle of cola, companies supersized the bottles, changing the eight-ounce Coke into the twenty-ounce size dispensed by most soda machines.

In addition to soft drinks, HFCS is added to many, if not most, of the products you'll find at the market. According to Pollan, of the sixty items on the McDonald's menu, forty-five contain HFCS. At the market, check the labels. You'll be surprised at how many products contain this product. For example, the HFCS is the third ingredient in ketchup, after tomatoes and vinegar.

SPOONFULS OF BROCCOLI

While I don't usually subscribe to conspiracy theories, it almost seems as if government and agribusiness—or industrial capitalism, as Pollan calls it—have conspired to undermine our health. American food is the cheapest in the world, relative to our income. One of the reasons food is cheap in this country is that government subsidizes farmers—but not all farmers. Our government subsidizes the farm products that are the likeliest to make us fat: corn, rice, soybeans, sugar, and wheat. These are the products that fatten hogs and beef and that make it possible for the food industry to offer inexpensive yet high-calorie products to consumers.

For every bushel they sell, farmers of these products receive money from the government. (The money, of course comes from our tax dollars, to the tune of about $25 billion a year.) If the market price of the crop drops, subsidy payments increase to make up the difference. Under these conditions, farmers naturally produce as much as they can. The result is an oversupply of the product. Oversupply makes the market price go down. Between 1996 and 2000, the price that food manufacturers had to pay for grain fell by nearly half. Frito-Lay could buy fifty-six pounds of corn for $2.56. (Corn farmers are the chief recipients of farm aid.) Without the subsidies, Doritos would cost more.

In contrast, farmers who grow vegetables and fruits do not receive government subsidies for producing these crops. As a consequence, we pay the market price for broccoli, artichokes, squash, plums, and apples. (The 2007 farm bill, still being debated at this writing, may provide support to growers of fruit and vegetables; support for grain growers will likely remain largely the same.) Between 1982 and September 2003, the consumer price of fresh fruits and vegetables increased 127 percent. But when we buy corn-fed beef, we reap the advantage of farm subsidies. A hamburger made with subsidized wheat (the bun) and subsidized beef is a bargain. Looked at another way, as the real price of fruits and vegetables increased by nearly 40 percent between 1985 and 2000, the real price of soft drinks declined by 23 percent. Clearly, Twinkies receive much more government support than tomatoes.

SOLO SNACKING

By creating and promoting easy-to-eat prepared foods, manufacturers make it easy to eat alone. Not so long ago, mealtime was a shared experience that occurred at set times. We sat down to a home-cooked meal with others and ate what was set before us. We not only started and stopped at specific times, but we adjusted our consumption to those who ate with us. It was socially unacceptable to eat in a greedy way. In fact, at one point in history, it was considered greedy to eat alone. Now, nearly half of Americans consume most of their meals away from home or on the go. We not only eat alone, we eat what we want, when we want, in the amounts we want. We eat all day long. We snack; we graze; we nosh. (And we're not usually snacking on a piece of broccoli; it's more likely to be a cheese Danish.) We eat standing up, walking, and driving. (No wonder cup holders in cars have grown along with us.)

To encourage our snacking habits, a major goal of the food industry is to elevate a food's "snackability"—make food easy to eat and conducive to mindless munching. Of course, most fruits and other healthful alternatives aren't as "snackable" as the prepared bite-sized morsels you'll find in a bag. According to a study conducted in 2000, apples, bananas, and oranges together accounted for less than 6 percent of the $80 billion snack-food market. (The good news is that apple growers are now competing with snack food makers by offering apples in easy-to-eat slices.) The latest thing in snackable foods are the hundred-calorie packs of Goldfish crackers, Wheat Thins, Reese's Snacksters, and the like. The idea, of course, is to reduce portion sizes and purportedly help people control their calorie intake. But the packages are highly profitable, having surpassed in 2007 the $20-million-a-year mark for the companies that produce them. Where the Goldfish are concerned, consumers pay $5.26 for the same amount of crackers—in nine three-quarter-ounce snack packs—as for one 6.6-ounce bag, which costs $1.99. Because the small packages are more portable, they make snacking even easier (not a good thing, as far as I'm concerned). Plus they create more trash.

In his book *Fat Politics*, J. Eric Oliver states, "It is the drinking and snacking more than anything else, that have been responsible for changing the way we eat." He cites studies showing that, in addition to an increase in Americans' soft drink consumption since 1978 (more than 130 percent), people today get nearly 25 percent of their daily calories from snacking, compared with 13 percent in the midseventies. Certainly, getting our calories from a bag of munchies is a lot less trouble than cooking.

OUNCES OF LEAVES

A hefty portion of those snackable foods is made from processed corn and wheat. In a January 27, 2007, *New York Times Magazine* article titled "Unhappy Meals," Michael Pollan notes, "[F]or reasons of economics, the food industry prefers to tease its myriad processed offerings from a tiny group of plant species, corn and soybeans chief among them. Today, a mere four crops accounts for two-thirds of the calories humans eat. When you consider that humankind has historically consumed some eighty thousand edible species, and that three thousand of these have been in widespread use, this represents a radical simplification of the food web." This matters, he says, because humans require "somewhere between 50 and 100 different chemical compounds and elements to be healthy. It's hard to believe that we can get everything we need from a diet consisting largely of processed corn, soybeans, wheat and rice."

Of course seeds (corn, rice, etc.), are easy to store and transport. By feeding the grain to animals, farmers can profitably transform, say, corn, into animal protein, and food-processing companies can transform them into Fritos, Cocoa Puffs, and Twinkies. (Twinkies, by the way, contain thirty-nine ingredients, most of which you wouldn't recognize. You can make a cake at home with six ingredients.) While corn, soybeans, wheat, and rice may be "perfectly suited to the needs of industrial capitalism," as Pollan says, they don't meet our nutritional need for micronutrients. "We're eating a lot more seeds and a lot fewer leaves" than our ancestors did.

TODDLERS ON THE BANDWAGON

If you were raised on snackable foods, it can be tough to get into the leaf-eating habit. The genie's out of the bottle, and we'll be hard pressed to put it back. It will be especially tough to change eating habits of young children, let alone our own. Since the late seventies and early eighties, when we began eating more processed, sugary foods, we've raised a generation of people habituated to such diets. Children's eating habits develop between ages two and three. Studies are showing that children are becoming accustomed to highly processed and refined foods, such as soft drinks, cookies, and sweetened cereals at increasingly younger ages. One study of four thousand children reports that 20 percent of children ages nineteen months to two years ate French fries *daily*; 23 percent of that age group drank sweetened beverages *daily*.

As a result, children in the one to two year age group were routinely taking in 30 percent more calories than they need.

Another study showed that the percentage of preschoolers who are overweight has risen from 5 percent in the 1970s to 13.5 percent in 2002. This weight gain correlates with an increase in consumption of fruit juices and soft drinks since 1977: a 42 percent increase in fruit juices and a 68 percent increase in soft drinks. In a study that monitored children's school meals, researchers found that on days when the kids ate fast food, they consumed an average of 187 more calories in one day than on days without fast food. Assuming that, on average, American kids eat a fast-food meal one out of every three days, they would stand to gain an extra six pounds of weight in a year. The statistics are bearing this out: in the sixties 4 percent of US children ages six to eleven were obese. In 2002, that number increased to 16 percent. As far as I'm concerned, obesity is less an issue here than the development of eating habits that rely on such a narrow range of nutrients.

POORER YET FATTER

According to the US Department of Health and Human Services, rates of childhood obesity rise as socioeconomic status falls. A trip to the supermarket shows us why: unhealthful food is significantly cheaper than healthful food. Adam Drewnowski, an obesity researcher at the University of Washington, proved this point. Giving himself a hypothetical dollar to spend, he discovered that he could buy 1,200 calories of cookies or potato chips but only 250 calories of carrots. His dollar bought 875 calories of soda but only 170 calories of orange juice. Processed (and subsidized) foods, which are typically located in the center of supermarkets, provide more energy in the form of calories than fresh food. If you don't have a lot of money, you get the most bang for your buck with junk food.

In Starr County, Texas, near the Mexican border, 59 percent of its children live below the poverty level. In this community, 50 percent of the boys in elementary school and 35 percent of the girls are over-weight or obese. Almost half the adults have Type 2 diabetes. Researchers studying this community postulate a number of reasons for the problem: easy access to inexpensive fast-food franchises, hot weather, and shortage of safe places for children to play outside. Genetics also plays a role. Some researchers speculate that many of the residents are descendents of the American Indian hunter-gatherers who are genetically predisposed to hang on to their fat.

Parental guilt may also contribute to the problem: parents feel guilty about denying their children food and children work the guilt angle to get what they want. As one school worker commented about the parents: "They love their children, but their children have them convinced that if they eat a healthy diet, they will starve." When school cafeterias tried revamping their menus from the popular sugar-coated cereal breakfasts and pizza lunches toward less caloric (and less popular foods), the children staged lunchroom protests.

INTERVENING IN THE SCHOOLS

The Starr County schools aren't the only ones being descended upon by interventionists. With all the hue and cry about childhood obesity, schools are increasingly the target of fat-reduction experiments and programs. As in Starr County, the programs are mostly ineffective—at least where weight loss is concerned. For example, one study, which lasted eight years and cost $20 million, followed 1,704 third graders in forty-one elementary schools. Leading the study was Benjamin Caballero, director of the Center for Nutrition at Johns Hopkins University's Bloomberg School of Public Health. Like the Starr County schools, the schools in this study were chosen because the students were mostly Native Americans—people at greater risk for obesity than the general population. Using random selection, some of the schools were left alone (the control group), while others were subjected to intensive intervention, which consisted of reduced-fat meals (both breakfast and lunch), healthful snacks, classroom lessons on choosing healthful foods, and educational programs for parents. The children also participated in physical exercises for an hour at least three times a week (the goal was an hour a day).

At the end of the study, the students could successfully parrot back their nutrition lessons, and the cooks could successfully prepare low-fat meals, but, as reported in the *American Journal of Clinical Nutrition*, "The intervention resulted in no significant reduction in percentage body fat. However, a significant reduction in the percentage of energy from fat was observed in the intervention schools." In other words, the kids ended up just as fat as the control group, even though they ate less fat.

A similar study, which followed 5,106 children from ninety-six schools in California, Louisiana, Michigan, and Texas, produced similar results: after three years, the kids in the schools that served reduced-fat food, received nutrition lessons, and underwent extra physical activities were no thinner than the control group. The obvious conclusion from both these studies—the second of which is described by the researchers as "the largest school-based randomized trial ever conducted"—is that weight is largely genetically determined.

After publication, both of these studies were ignored—and continue to be ignored. Because of increasing social pressure, schools will continue to be the targets of "interventions." As Gina Kolata tells us in her book *Rethinking Thin*, the National Institutes of Health, which sponsored the two large studies, is "acting as though the studies never existed, starting a new program called 'We Can' to involve parents and schools in changing children's diet and adding more exercise into their day."

This may not be all bad. Exercise is a good thing. Removing soft drinks from schools is a good thing. Vegetables and fruits are good things. But the focus of "We Can" is clearly on getting children to lose weight, equating being thin with being healthy. And, predictably, their nutritional information lists fat as a "WHOA" food (food to be avoided). Even though the huge studies on school children showed that reducing their fat intake did not result in thinner kids, the "We Can" program still casts fat in its old demonic role.

PROGRAMMING BABIES

Perhaps in the future we won't have to worry about obese children. Scientists at the Clore Laboratory, University of Buckingham in the UK are studying ways to program infants to be less prone to obesity. A new field of research, called developmental programming, maintains that, during the period that brains and other organs are still being fine tuned, it's possible to affect the tuning. According to Michael Cawthorne, director of metabolic research at the laboratory, it may be possible to program babies' metabolisms in a way that provides permanent resistance to obesity. In this case, the programming would come in the form of a baby formula containing leptin, the hormone produced by fat cells that helps regulate appetite (see chapter 6). In their experiments the researchers discovered that by giving injections of leptin to female rats during pregnancy and lactation, their offspring were resistant to obesity. Apparently the injected leptin influences the developing brain's set points for hunger suppression and calorie burning. The scientists postulate that, by feeding leptin-laced "anti-obesity formula" to infants, the same will happen to them. Maybe so; but I'm not so sure it's a good idea to mess around with an infant's developing brain just so they'll be thin adults.

OLD-FASHIONED EATING

Though the kids will protest, the best course of action is to return to old-fashioned meals, the kind you have to cook—or at least prepare—yourself. Essentially this means avoiding what Pollan calls "industrial food"—food that's been processed beyond recognition. Returning to traditional eating does not require such a giant leap backward. Our modern diet of processed and refined foods is a rather recent development, especially if you take the long view. A hundred thousand generations of people lived as hunter-gatherers; five hundred generations depended on agriculture; ten generations have lived since the start of the industrial age; and only two generations have grown up with highly processed foods. Our bodies are now virtually the same as they were forty thousand years ago, but our eating habits bear little resemblance to that of our early ancestors.

If you wanted to get really traditional, you could look at what our hunter-gatherer ancestors ate about forty thousand years ago. You may not want to emulate their diet—which included lots of grit, dirt, and ashes—or take up hunting and gathering, but it's worth considering the sort of food we were programmed to eat. We're designed to subsist on a diet consisting mostly of plant parts: nuts, roots, leaves, and fruits. Depending on the season and location, our ancestors also ate meat, including fish and shellfish; scientists disagree on the amount. Meat or no meat, the human digestive system, which is essentially the same as a chimpanzee's, was designed for processing fiber. In fact, the early human's diet was similar to that of the chimpanzee.

FINDING REAL FOOD

Ideally we should be eating food that's been raised to approximate the way nature intended. Following this advice may not be easy. For example, you'd want to eat grass-fed beef instead of the corn-fed beef that's available at your local supermarket. Grass-fed beef more closely resembles the wild game we were programmed to eat. Cattle are ruminants—that is, they possess a rumen, a forty-five-gallon "fermentation tank" in which resident bacteria convert cellulose into protein and fat. In our modern feedlots, however, cattle are fattened on corn, a feed unsuited to their natural diet and one that makes them ill with bloat, acidosis, ulcers, and a whole host of other diseases. One veterinarian noted that between 15 and 30 percent of feedlot cows have abscessed livers; in some places, the number runs as high as 70 percent. To keep the cows healthy (enough) they are given antibiotics. In fact, most of the antibiotics sold in America end up in animal feed.

While grass is a healthier diet for cattle, grass-fed beef is also healthier for us. It contains more omega-3 fatty acids, beta-carotene, vitamin E, conjugated linoleic acid (CLA), and folic acid than corn-fed beef. Plus, you can probably count on grass-fed beef to be raised without antibiotics, hormones, or rendered animal by-products. Like-wise, milk and eggs from pastured animals also contain higher levels of fat-soluble vitamins as well as omega-3 fatty acids—all derived from the green plants such animals are permitted to eat. You can go so far as to drink raw milk to get the heat-sensitive folic acid, some of which is destroyed by pasteurization. Unfortunately these healthful alternatives to industrial foods are not typically available at your supermarket. You'll have to go to the health-food store and you'll also have to pay more. You can also find sources at www.eatwild.com. Or you can raise your own. (I've tried that. It's not easy.)

AN EVOLUTION SOLUTION

Perhaps someday evolution will come to the rescue. Scientists tell us that evolution didn't grind to a halt in the distant past. People have continued to adapt to the pressures of new diseases, climates, and diets, and natural selection has continued to favor the best adapted. By looking at the characteristics and genomes of various peoples around the globe, researchers can see evidence of continued evolution. The ability to digest lactose is a case in point. Lactase, the enzyme that digests lactose, which is the principal sugar in milk, is usually switched off after weaning. But about five thousand years ago, among the cattle-herding people of northern Europe, natural selection began to favor those for whom a genetic change enabled them to digest lac-tose in adulthood, a significant nutritional benefit. It turns out that lac-tose tolerance evolved independently four times, according to researchers at the University of Maryland, who tested forty-three ethnic groups in East Africa. The researchers found three separate mutations, all different from the European one, that keep the lactase gene switched on in adulthood. In other words, through natural selection, the people who could digest lactose were better adapted to survive than those who couldn't.

Who knows? Like those who became adapted to lactose, maybe we can adapt to highly refined processed food, it being the most abundant in our current environment. Those who thrive on it will have the competitive advantage, especially as prepared foods proliferate and fresh foods disappear from the grocery shelves. Until then, we'll just have to live with the genes we've got and be thankful our ancestors were fat enough to survive.

Appendix A.

CHEMICAL SUBSTANCES PRODUCED OR ACTIVATED BY FAT CELLS

fat cells produce the following substances (mostly hormones) and release them into the bloodstream:

Name	What it does	Comments
Leptin	Suppresses or increases appetite; as fat is stored in cells, leptin is released to suppress appetite.	Some obese people seem resistant to leptin's effects.
Adiponectin	Stimulates the body to burn fatty acids as fuel; reduces vascular inflammation; increases cells' sensitivity to insulin.	Adiponectin is beneficial; low levels increase risk for Type 2 diabetes and heart disease; the more fat, the *less* adiponectin is produced.
Resisten	Makes cells resistant to insulin; encourages the liver to convert fatty acids into glucose.	High levels can increase the risk of Type 2 diabetes.
Plasminogen activator inhibitor-1	Blocks the body's clot-busting agents.	High levels can increase risk of heart attack and stroke.

Angiotensinogen	Constricts blood vessels.	High levels can lead to high blood pressure.
PA-I-1	Blocks the body's clot-busting agent.	High levels can increase risk of heart attack and stroke.
IL-6	Contributes to low-grade inflammation.	May be implicated in heart disease, Type 2 diabetes, and certain types of cancer.
Tumor necrosis factor alpha	Interferes with the operation of insulin.	High levels can increase the risk of Type 2 diabetes.
Estrogen	Multiple effects on development and regulation of female reproductive system.	Is linked to breast cancer.
Macrophages	Immune system cells that produce inflammatory compounds.	In cases of obesity, macrophages can invade fat cells, acting as if the fat itself were an invading organism.
Cortisol	Encourages the deposition of fat in the abdomen.	Is produced in the adrenal glands, but fat cells can activate an inactive version of this hormone.

Sources: Denise Grady, "Fat—The Secret Life of a Potent Cell," *New York Times*, July 6, 2004; Ann Underwood and Jerry Adler, "What You Don't Know about Fat," *Newsweek*, August 23, 2004.

Appendix B.

DETERMINING YOUR BODY FAT USING GIRTH MEASUREMENTS

To find out what percentage of your body consists of fat compared to other tissues, use the following techniques to measure your various body parts, calculate the measurements, and interpret the results (adapted from *The Ultimate Fit or Fat*, by Covert Bailey [New York: Houghton Mifflin, 2000]). Use a cloth tape to perform the measurements as follows:

Women:

Hips: Standing with your feet about four inches apart, measure your hips at the largest circumference.

Thigh: Standing with your feet about twelve inches apart, measure the upper thigh at the widest part.

Calf: Standing with your weight evenly distributed on both feet, measure your calf at the widest part.

Wrist: Measure just above the bony protuberance toward your hand.

Men:

Waist: Relax your gut before measuring.

Hips: Standing with your feet about four inches apart, measure the largest circumference.

Forearm: Clench your fist to flex your forearm, then measure at the widest part.

Wrist: Measure just above the bony protuberance toward your hand.

Note: Measure your dominant arms and legs.

Use the following formulas to calculate percent of body fat based on your measurements:

Women 30 years and younger:

Hips + (.80 × thigh) – (2 × calf) – wrist = % body fat

That is, multiply .80 times your thigh measurement and add that amount to your hip measurement. Double your calf measurement and subtract that from the calculation you just did. From that amount, subtract your wrist measurement to get your body fat percentage.

Women over 30:

Hips + thigh – (2 × calf) – wrist = % body fat

Men 30 years and younger:

Waist + (½ hips) – (3 × forearm) – wrist = % body fat

Men over 30

Waist + (½ hips) – (2.7 × forearm) – wrist = % body fat

Interpreting the results: Typically the tape measure method is usually within 2 percent of the results obtained from underwater weighing. However, in some cases, results can vary by 3 to 5 percent, depending on the amount of intermuscular fat, which the tape measure method doesn't calculate. Extremely fit people may get higher percentages using the tape measure method because they may have less intermuscular fat; people who may be skinny but are not very fit may get lower percentages because they may have more than the usual amount intermuscular fat.

Appendix C.

FAT CONTENT OF FOODS

here is a list of selected common food items together with their total fat, saturated, monounsaturated, and polyunsaturated fatty acids, cholesterol, lauric acid, omega-3 and omega-6 fatty acids. Omega-3 and omega-6 are both polyunsaturated fatty acids. Lauric acid is included because of its health benefits. It's a medium-chain saturated fatty acid known to the pharmaceutical industry for its antimicrobial properties, meaning that it inhibits the growth of bacteria, viruses, and fungi. The source of this information is *Know Your Fats,* by Dr. Mary G. Enig. In the table, "g" stands for grams and "mg" stands for milligrams.

Food	Calories	Total fat (g)	Sat. fat (g)	Mono fat (g)	Poly fat (g)	Chol. (mg)	Lauric acid (g)	Omega-3 (g)	Omega-6 (g)
Dairy									
Butter, 1 tbsp.	102	11.5	7.2	3.3	0.04	31	0.32	0.17	0.26
Whole (3.7% fat) milk, 8 oz.	157	8.9	5.6	0.3	0.2	35	0.25	0.13	0.20
2% fat milk, 8 oz.	121	4.7	2.9	1.4	0.2	18	0.13	0.07	0.10
1% fat milk, 8 oz.	102	2.6	1.6	0.8	0.1	10	0.07	0.04	0.06
Skim milk, 8 oz.	86	0.4	0.3	0.1	0.0	4	0.01	0.00	0.01
Whole milk yogurt, plain, 1 cup	150	8.0	5.1	2.2	0.2	31	0.27	0.07	0.16
Low-fat yogurt, 1 cup	155	3.8	2.5	1.0	0.1	15	0.13	0.03	0.08
Cultured sour cream, 1 tbs.	31	3.0	1.9	0.9	0.1	6	0.08	0.04	0.07
Heavy whipping cream, 1 tbs.	51	5.5	3.4	1.6	0.2	20	0.15	0.08	0.12
Half & half cream, 1 tbs.	20	1.7	1.1	0.5	0.1	6	0.05	0.02	0.04
Cheese									
Asiago, 2 oz.	213	15.6	10.1	4.1	0.6	52	0.30	0.20	0.35
Blue, 2 oz.	200	16.3	10.6	4.4	0.5	43	0.28	0.15	0.30
Cheddar, 2 oz.	228	18.8	12.0	5.3	0.5	60	0.30	0.21	0.33
Cottage cheese, creamed, 2 oz.	58	2.3	1.7	0.7	.05	8	—	0	0
Cottage cheese, 1% low fat, 2 oz.	45	1.2	0.5	.15	.01	2.5	—	0	0
Cream cheese, 2 oz.	198	19.8	12.5	5.6	0.7	62	0.26	0.28	0.44
Edam, 2 oz.	202	15.8	10.0	4.6	0.4	51	0.28	0.14	0.24
Gorgonzola, 2 oz.	204	17.7	10.8	4.9	1.0	49	0.73	0.52	0.46
Goat, semi-soft, 2 oz.	206	16.9	11.7	3.9	0.4	45	0.75	0	0.40

Food									
Gruyere, 2 oz.	234	18.3	10.7	5.7	1.0	62	0.52	0.24	0.74
Monterey Jack, 2 oz.	211	17.2	10.8	5.0	0.5	50	0.23	0.15	0.36
Mozzarella, part skim, 2 oz.	159	9.7	6.2	2.7	0.3	31	0.10	0.08	0.20
Mozzarella, whole milk, 2 oz.	159	12.3	7.5	3.7	0.4	44	0.39	0.21	0.22
Parmesan, grated, 2 oz.	259	17.0	10.8	4.9	0.4	45	0.56	0.20	0.18
Ricotta, part skim milk, 2 oz.	78	4.5	2.8	1.3	0.1	17	0.06	0.04	0.11
Ricotta, whole milk, 2 oz.	99	7.4	4.7	2.0	0.2	29	0.09	0.06	0.15
Roquefort, 2 oz.	209	17.4	10.9	4.8	0.7	51	0.74	0.40	0.35
String, 2 oz.	144	9.0	5.7	2.6	0.3	34	0.09	0.08	0.19
Swiss, shredded, 2 oz.	213	15.6	10.1	4.1	0.6	52	0.30	0.20	0.35
Yogurt, 2 oz.	43	0.1	0.1	0.0	0.0	2	—	—	—
Fruit/Seed Oils									
Canola oil, 1 tbsp.	120	13.6	1.0	8.0	4.0	0	0	1.3	2.8
Cocoa butter, 1 tbsp.	120	13.6	8.1	4.5	0.4	0	0	0.01	0.38
Coconut oil, 1 tbsp.	117	13.6	11.8	0.8	0.2	0	6.01	0	0.25
Corn oil, 1 tbsp.	120	13.6	1.7	3.3	8.0	0	0	0.11	7.90
Cottonseed oil, 1 tbsp.	120	13.6	3.5	2.4	7.1	0	0	0.03	7.03
Extra virgin olive oil, 1 tbsp.	126	14	2.0	10.8	1.3	0	—	0.10	1.12
Flaxseed oil, 1 tbsp.	120	13.5	1.3	2.3	9.3	—	0	7.52	1.82
Palm oil, 1 tbsp.	117	13.6	11.1	1.6	0.2	0	6.4	0	0.22
Peanut oil, 1 tbsp.	119	13.5	2.3	6.2	4.3	0	0	0	4.32
Safflower oil, high linoleic, 1 tbsp.	120	13.6	1.2	1.6	10.2	—	0	0.05	10.10
Safflower oil, high oleic, 1 tbsp.	120.	13.6	0.8	10.2	1.9	0	0	0	1.93
Sesame oil, 1 tbsp.	120	13.6	1.9	5.4	5.7	0	0	0.04	5.63

Food	Calories	Total fat (g)	Sat. fat (g)	Mono fat (g)	Poly fat (g)	Chol. (mg)	Lauric acid (g)	Omega-3 (g)	Omega-6 (g)
Soybean oil (e.g., Wesson), 1 tbsp.	120	13.6	2.0	3.2	7.9	0	0	0.93	6.95
Sunflower oil, high linoleic, 1 tbsp.	120	13.6	1.4	2.7	9.0	0	0	0	8.95
Sunflower oil, high oleic, 1 tbsp.	124	14	1.4	11.7	0.5	0	0	0.03	0.50
Meats, including poultry									
Beef steak, 3.5 oz.	205	9.3	3.6	3.8	0.3	75	0.01	0.02	0.28
Beef roast, bottom round, 3.5 oz.	273	16.8	6.3	7.3	0.6	95	0.04	0.17	0.47
Beef chuck, 3.5 oz.	214	8.2	3.0	3.4	0.3	100	0.01	0.02	0.30
Veal loin chop, braised, 3.5 oz.	224	9.1	2.5	3.2	0.8	124	0.02	0.04	0.77
Lamb loin chop, broiled, 3.5 oz.	314	22.9	9.8	9.6	1.7	99	0.10	0.35	1.32
Pork roast, center loin, 3.5 oz.	198	8.9	3.3	4.0	0.7	78	0.01	0.02	0.65
Beef liver, fried, 3.5 oz.	215	8.0	2.7	1.6	1.7	478	0	0.29	1.33
Chicken breast, roasted, 3.5 oz.	196	7.7	2.2	3.0	1.7	83	0.01	0.10	1.48
Chicken thigh, roasted, 3.5 oz.	245	15.4	4.3	6.1	3.4	92	0.03	0.20	3.11
Turkey, white, roasted, 3.5 oz.	196	8.3	2.3	2.8	2.0	75	0.01	0.13	1.80
Turkey, dark, roasted, 3.5 oz.	219	11.4	3.5	3.6	3.1	88	0.01	0.19	2.84
Goose, roasted, 3.5 oz.	303	21.7	6.8	10.2	2.5	90	0.04	0.18	2.24
Duck, roasted, 3.5 oz.	334	28.2	9.6	12.8	3.6	83	0.04	0.29	3.34
Chicken livers, simmered, 3.5 oz.	156	5.4	1.8	1.3	0.9	626	0	0.11	0.73
Goose liver, raw, 3.5 oz.	132	4.3	1.6	0.8	0.3	511	0	0.01	0.25
Venison, roasted, 3.5 oz.	157	3.2	1.2	0.9	0.6	111	0	0.09	0.53
Beefalo, roasted, 3.5 oz.	186	6.3	2.7	2.7	0.2	58	0.01	0.06	0.14

Beef salami, 2 oz.	148	11.7	5.1	5.4	0.6	37	0.03	0.13	0.45
Pork salami, dry, hard, 2 oz.	231	19.1	6.7	9.1	2.1	45	0	0.16	1.94
Beef bologna, 2 oz.	177	16.2	6.9	8.0	0.6	33	0.02	0.14	0.48
Pork bologna, 2 oz.	140	11.3	3.9	5.6	1.2	33	0.01	0.16	1.04
Liverwurst, pork, 2 oz.	1.85	16.2	6.0	7.5	1.5	90	0	0.08	1.39
Italian pork sausage, 2 oz.	183	14.6	5.2	6.8	1.9	44	0.03	0.25	1.61
Polish sausage, pork, 2 oz.	185	16.3	5.9	7.7	1.7	40	0.04	0.16	1.58
Pepperoni sausage, 2 oz.	282	24.9	9.1	12.0	2.5	45	0	0.23	2.20
Bacon, pork, cooked, 2 oz.	327	27.9	9.9	13.5	3.3	48	0.04	0.45	2.85
Canadian bacon, 2 oz.	105	4.8	1.6	2.3	0.5	33	0.01	0.07	0.40
Fin fish, baked or broiled									
Atlantic cod, 3.5 oz.	104	0.8	0.2	0.1	0.3	55	0	0.16	0.03
Sole/flounder, 3.5 oz.	116	1.5	0.4	0.2	0.6	67	0	0.51	0.06
Haddock filet, 3.5 oz.	111	0.9	0.2	0.1	0.3	73	0	0.24	0.04
Atlantic ocean perch, 3.5 oz.	120	2.1	0.3	0.8	0.5	54	0	0.44	0.04
Walleye pike, 3.5 oz.	118	1.5	0.3	0.4	0.6	109	0	0.41	0.11
Sea bass fillet, 3.5 oz.	123	2.5	0.6	0.5	0.9	53	0	.76	0.03
Snapper fillet, 3.5 oz.	127	1.7	0.4	0.3	0.6	47	0	0.32	0.07
Fresh water bass, 3.5 oz.	145	4.7	1.0	1.8	1.3	86	0	0.90	0.29
Farm channel catfish, 3.5 oz.	151	8.0	1.8	4.1	1.4	64	0	0.24	1.01
Halibut fillet, 3.5 oz.	130	2.9	0.4	1.0	0.9	41	0	0.55	0.22
Herring, pickled, one, 20 g	52	3.6	0.5	2.4	0.3	3	0	0.28	0.04
Spanish mackerel fillet, 3.5 oz.	157	6.3	1.8	2.1	1.8	72	0.01	1.33	0.26
Atlantic salmon fillet, wild, 3.5 oz.	181	8.1	1.2	2.7	3.2	70	—	2.20	0.56

Food	Calories	Total fat (g)	Sat. fat (g)	Mono fat (g)	Poly fat (g)	Chol. (mg)	Lauric acid (g)	Omega-3 (g)	Omega-6 (g)
Sockeye salmon, canned, 3.5 oz.	152	7.3	1.6	3.1	1.9	44	0	1.24	0.45
Swordfish, 3.5 oz.	154	5.1	1.4	2.0	1.2	50	0	1.08	0.12
Fresh bluefin tuna, 3.5 oz.	183	6.2	1.6	2.0	1.8	49	0	1.49	0.12
White tuna, in water, 172 g	220	5.1	1.4	1.3	1.9	72	0	1.60	0.18
Whitefish fillet, 3.5 oz.	171	7.4	1.2	2.5	2.7	76	0	1.84	0.63
Atlantic mackerel, 3.5 oz.	260	17.7	4.1	7.0	4.3	74	0.02	1.30	0.20
Sockeye salmon, 3.5 oz.	214	10.9	1.9	5.2	2.4	86	0	1.28	0.14
Chinook salmon, 3.5 oz.	229	13.3	3.2	5.7	2.6	84	0	1.83	0.33
Farmed Atlantic salmon, 3.5 oz.	204	12.3	2.5	4.4	4.4	62	0	2.14	1.83
Crustaceans and mollusks, steamed or boiled									
Alaskan king crab leg, 134 g	130	2.1	0.2	0.2	0.7	71	0	0.57	0.08
Blue crab, 48 g	49	0.8	0.1	0.1	0.3	48	0	0.24	0.05
Whole Dungeness Crab, 127 g	140	1.6	0.2	0.3	0.5	96	0	0.50	0
Blue crab cakes, 60 g	93	4.5	0.9	1.7	1.4	90	0	0.32	1.00
Northern lobster, 3.5 oz.	97	0.6	0.1	0.2	0.1	71	0	0.08	0
Octopus, 3.5 oz.	163	2.1	0.4	0.3	0.5	95	0	0.31	0.10
Prawns/shrimp, 3.5 oz.	98	1.1	0.3	0.2	0.4	193	0	0.32	0.09
Large clams, 3.5 oz.	147	1.9	0.2	0.2	0.5	66	0	0.29	0.11
Eastern oysters, wild, 3.5 oz.	136	4.9	1.5	0.6	1.9	104	0	1.21	0.27
Pacific oysters, 3.5 oz.	162	4.6	1.0	0.7	1.8	99	0	1.43	0.14
Scallops, 3.5 oz.	106	3.1	0.5	1.1	1.0	32	0	0.77	0.10

Seeds and nuts

Flax seeds, 2 tbsp.	95	6.9	0.6	1.3	4.3	0	0	3.51	0.84
Pumpkin seeds, dry, 2 tbsp.	93	7.9	1.5	2.5	3.6	0	0.01	0.03	3.57
Almonds, dry, 2 tbsp.	105	9.3	0.9	6.0	1.9	0	0	0.07	1.86
Brazil nuts, dry, 2 tbsp.	115	11.6	2.8	4.0	4.2	0	0	0.01	4.16
Cashew nuts, dry roasted, 2 tbsp.	98	7.9	1.6	4.7	1.3	0	0.13	0.03	1.31
Coconut, fresh, shredded, ½ cup	142	13.4	12.0	0.6	.01	0	5.96	0	0.15
Coconut, dried, unsweet, ½ cup	257	25.2	22.5	1.1	0.3	0	11.15	0	0.28
Coconut milk, canned, ½ cup	223	24.1	21.6	1.0	0.3	0	10.69	0	0.26
Hazelnuts, chopped, 2 tbsp.	91	9.0	0.7	7.1	0.9	0	0	0.02	0.84
Macadamia nuts, dried, 2 tbsp.	118	12.3	1.8	9.7	0.2	0	0	0	0.21
Pecans, dry, 2 tbsp.	99	10.0	0.8	6.3	2.5	0	0	0.10	2.38
Pine nuts, dry, 2 tbsp.	96	8.6	1.3	3.2	3.6	0	0	0.11	3.51
Pistachio nuts, dry, 2 tbsp.	92	7.7	1.0	5.2	1.2	0	0	0.04	1.12
Black walnuts, chopped, 2 tbsp.	95	8.8	0.6	2.0	5.9	0	0	0.52	5.23
English walnuts, chopped, 2 tbsp.	96	9.3	0.9	2.1	5.9	0	0	1.02	4.77

Nut butters

Almond butter, 2 tbsp.	203	18.9	1.8	12.3	4.0	0	0	0.14	3.81
Cashew butter, 2 tbsp.	188	15.8	3.1	9.3	2.7	0	0.26	0.05	2.61
Peanut butter, 2 tbsp.	190	16.3	3.3	7.8	4.4	0	0.01	0.02	4.38
Sesame butter (tahini), 2 tbsp.	190	16.3	2.3	6.1	7.1	0	0	0.12	7.01

Misc. vegetables

Tomatoes, boiled, 1 cup	65	1.0	0.1	0.2	0.4	0	0	0.01	0.39
Soybeans, green, boiled, ½ cup	127	5.8	0.7	1.1	2.7	0	0	0.32	2.39

Food	Calories	Total fat (g)	Sat. fat (g)	Mono fat (g)	Poly fat (g)	Chol. (mg)	Lauric acid (g)	Omega-3 (g)	Omega-6 (g)
Avocado, Calif., pureed, 1 cup	407	39.8	6.0	25.8	4.7	0	0	0.26	4.43
Avocado, Flor., pureed, 1 cup	258	20.4	4.0	11.2	3.4	0	0	0.20	3.21

REFERENCES

CHAPTER 1. GOOD THINGS ABOUT FAT

Stored Energy

Biochemistry II lecture on the Web (lecture 24), University of Texas at Dallas, Spring, 2002, http://216.239.53.104/search?q=cache:yhuatdVgu TgJ:nsm1.utdalas.edu/bioweb.Lectures (accessed March 2, 2004).

Mary Enig, *Know Your Fats: The Complete Primer for Understanding the Nutrition of Fats, Oils, and Cholesterol* (Silver Spring, MD: Bethesda Press, 2000), p. 75.

Lecture 2, "Dietary Fats, Body Fats and Blood Lipids," UCLA Center for Human Nutrition, http://www.cellinteractive.com/ucla/nutrition_101/ phys_lect2.html (accessed March 3, 2004).

Caroline M. Pond, *The Fats of Life* (Cambridge: Cambridge University Press, 1998), pp. 61, 62, 233.

Sharman Apt Russel, *Hunger: An Unnatural History* (New York: Perseus, 2005), pp. 7–8. [Source for the bit about the man who fasted for more than a year.]

Helen Ruppel Shell, *The Hungry Gene: The Science of Fat and the Future of Thin* (New York: Atlantic Monthly Press, 2002), pp. 34, 50.

Kathy Wollard, "Fat: A Need and Friend Indeed," Newsday.com, May 11,

2004, http://www.newsday.com/news/health/ny-hshow113797009may 11,0,4142701 (accessed May 11, 2004).

Hormone Production

Daniel Q. Haney, "Fat Cells Actively Breed Disease," *San Jose Mercury News*, May 11, 2004.
Pond, *The Fats of Life*, pp. 35, 153–54.

Cell Membrane Material

John Whitfield, "The Need for Fat," Horizon Symposia proceedings, October 2004, organized by Nature Publishing Group, http://www.nature.com/horizon/livingfrontier/background/fat.html (accessed July 10, 2007).

Padding

Pond, *The Fats of Life*, pp. 207, 211–13.

Insulation

Pond, *The Fats of Life*, pp. 196, 197.

The Immune System

C. M. Pond, "Adipose Tissue, the Immune System and Exercise Fatigue: How Activated Lymphocytes Compete for Lipids," *Biochemical Society Transactions* 30, part 2 (2002): 270–75.
———, *The Fats of Life*, pp. 279–85.
"Sudden Removal of Fat Impairs Immune Function in Rodents, Biologists Find," news release from Indiana University Media Relations, http://newsinfo.iu.edu/news/page/normal/865.html (accessed April 2, 2004). The research report was published in the May 7, 2003, issue of the *Proceedings of the Royal Society B*.

Protection from Toxic Substances

Andrew Weil, *Healthy Aging* (New York: Alfred A. Knopf, 2002), pp. 179–80.

Sexual Attraction

Helen Fisher, "Highly Intelligent Design," *New York Times Book Review*, August 28, 2005, reviewing Desmond Morris's *The Naked Woman—A Study of the Female Body.*
Pond, *The Fats of Life*, pp. 42–45, 164, 244–47.

Fat-Free People

Abigail Zuger, "Too Little Fat Can Be as Bad as Too Much," *New York Times*, July 6, 2004.

CHAPTER 2. WHAT IS FAT, ANYWAY?

Triglycerides Deconstructed

AS Guru Biology, BBC Education, April 2004, http://www.bbc.co.uk/education/asguru/biology/02biologicalmoecules/03/lipids (accessed April 20, 2004).
Biochemistry II lecture, number 24, University of Texas at Dallas.
Enig, *Know Your Fats*, pp. 9–11.
Pond, *The Fats of Life*, pp. 8–15.

Fatty Acids Deconstructed

AS Guru Biology, BBC Education, April 2004.
Biochemistry II lecture 24, University of Texas at Dallas.
Enig, *Know Your Fats*, pp. 11–15.
Pond, *The Fats of Life*, pp. 8–15.

Saturated, Monounsaturated, and Polyunsaturated Fatty Acids

Enig, *Know Your Fats*, pp. 23, 33–38.
Pond, *The Fats of Life*, pp. 77–85.

A Separate Case: *Trans* Fatty Acids

Enig, *Know Your Fats*, pp. 38–40.
Pond, *The Fats of Life*, pp. 97–101.

CHAPTER 3. HOW YOUR BODY DIGESTS AND USES FAT

First, the Stomach

"Dietary Fats, Body Fats and Blood Lipids," UCLA Center for Human Nutrition, lecture 2.

Next, the Intestines

"Absorption of Lipids," Hypertexts for Biomedical Sciences, Colorado State University, 2003, http://arbl.cvmbs.colostate.edu/hbooks/pathphys/digestion/smallgut/absorb_lipids.html (accessed March 18, 2004).
Lecture 2, "Dietary Fats, Body Fats and Blood Lipids," UCLA Center for Human Nutrition.
"Lipid Digestion and Lipoproteins," Medical Biochemistry, Indiana State University, http://web.indstate.edu/thcme/mwking/lipoproteins.html (accessed March 22, 2004).
Sherwin B. Nuland, *How We Live* (New York: Vintage Books, 1998), p. 297.
Pond, *The Fats of Life*, pp. 114–15.

Getting through Intestinal Walls

"Absorption of Lipids," Colorado State University.
Michael D. Gershon, *The Second Brain* (New York: HarperPerennial, 1998), pp. 116, 140–43.
"Intestinal Uptake of Lipids," Indiana State University, Medical Biochemistry, http://web.indstate.edu/thcme/mwking/lipoproteins.html/uptake (accessed March 22, 2004).
Lecture 2, "Dietary Fats, Body Fats and Blood Lipids," UCLA Center for Human Nutrition.
"Lipids Digestion, Absorption, and Transport," lecture outlines, Texas A&M University, http://animalscience.tamu.edu/nutr/202s/LectureOutlines/lipids2.html (accessed March 2, 2004).
Pond, *The Fats of Life*, pp. 117–18.

Transportation to Tissues

"Absorption of Lipids," Colorado State University.
Gershon, *The Second Brain*, p. 143.
John Ross, "The Role of the Lymphatic System in Fat Absorption and Transport," May 1, 1999, http://www.jdaross.mcmail.com/lymphatics3.htm (accessed March 22, 2004).

Unabsorbed Fats: The Laxative Effect

Gershon, *The Second Brain*, p. 144.
Pond, *The Fats of Life*, p. 118.

Substances along for the Ride

Pond, *The Fats of Life*, pp. 25, 116, 117.

Digestive Speed

Gershon, *The Second Brain*, p. 144.
Pond, *The Fats of Life*, p. 115.

The Food-to-Fat Conversion: When You Eat More Than You Need

Pond, *The Fats of Life*, p. 151.

The Food-to-Energy Conversion: When You Eat Less Than You Need

Pond, *The Fats of Life*, p. 132.

Fat as Fuel for Sustained Exercise

A. E. Jeukendrup et al., "Negative Fat Balance in Weight Stable Physically Active Humans on a Low-Fat Diet," *Journal of Physiology* 523 (December 1999): 223.

David R. Pendergast, John J. Leddy, and Jaya T. Venkatraman, "A Perspective on Fat Intake in Athletes," *Journal of the American College of Nutrition* 19, no. 3 (2000): 345–50.

Pond, *The Fats of Life*, pp. 141, 142.

CHAPTER 4. WHAT YOUR FAT CELLS DO

The Anatomy of a Fat Cell

Craig C. Freudenrich, "How Fat Cells Work," http://www.howstuffworks.com/fat-cell/htm (accessed September 6, 2005).

Denise Grady, "Fat—The Secret Life of a Potent Cell," *New York Times*, July 6, 2004, p. D1.

Shell, *The Hungry Gene*, p. 50.

Ann Underwood and Jerry Adler, "What You Don't Know about Fat," *Newsweek*, August 23, 2004, pp. 39–45.

Fat Cell Numbers

Pond, *The Fats of Life*, pp. 32, 51, 74.
Shell, *The Hungry Gene*, pp. 50, 51.
Underwood and Adler, "What You Don't Know about Fat," p. 42.

Fat as a Hormone Producer

Grady, "Fat—The Secret Life of a Potent Cell."
Pond, *The Fats of Life*, pp. 37, 153–56.
Underwood and Adler, "What You Don't Know about Fat," pp. 43, 44.

Hormones Worth Special Mention

S. J. Cleland and J. L. Reid, "The Role of Adiponectin in Coronary Disease and the Insulin Resistance Syndrome," *Servier* (December 8, 2003), http://www.servier.com/pro/cardiologie/pdfs (accessed August 9, 2004).
Maria G. Essig, "Low Adiponectin Levels Related to Development of Type 2 Diabetes," *Endocrinology* (September 23, 2002), http://www.obgyn.net/newsheadlines/headlinemedicalnews-endocrinology-20020923-4asp (accessed August 9, 2004).
Tohru Funahashi, "Adiponectin and the Metabolic Syndrome," conference presentation, March 29, 2003, http://www.cmeondiabetes.com/pub/adiponectin.an.the.metabolic.syndrome.php (accessed August 9, 2004).
Grady, "Fat—The Secret Life of a Potent Cell."
"Leptin 'Rewires' Neurons in Part of Brain That Regulate Feeding Behavior," Howard Hughes Medical Institute, April 2, 2004, http://diabetes.about.com/cs/newswire/a/blnobesewire404_p.htm (accessed August 9, 2004).
Underwood and Adler, "What You Don't Know about Fat," pp. 43, 44.

Fat Depots (Storage Areas)

Grady, "Fat—The Secret Life of a Potent Cell."
Eli Ipp, "Where's the Fat?" October 1999, http://www.thedoctorwillseeyounow.com/articles/nutrition/fatdistribution_1 (accessed January 18, 2005).
Pond, *The Fats of Life*, pp. 38–52, 156.

Apples, Pears, and Other Fat Storage Arrangements

Alicia di Rado, "The Skinny on Fat," *USC Health*, Winter 2004.

Gary Taubes, *Good Calories, Bad Calories* (New York: Alfred A. Knopf, 2007), pp. 397–98.

Underwood and Adler, "What You Don't Know about Fat," p. 45.

"Where's the Fat?" *Your Health* newsletter, Sutter Santa Cruz, Fall 2003.

Visceral Fat versus Subcutaneous Fat

Jane E. Allen, "Belly Fat Carries Greater Disease Risk Than Fat Elsewhere," *Los Angeles Times*, March 23, 2004, http://www.iconocast.com/H/Health3_News09_04/Health9F.htm (accessed May 11, 2004).

Grady, "Fat—The Secret Life of a Potent Cell."

Visceral Fat and Diabetes

Michael I. Goran, Richard N. Bergman, and Barbara A. Gower, "Influence of Total vs. Visceral Fat on Insulin Action and Secretion in African American and White Children," *Obesity Research* 9, no. 8 (August 2001): 423–31.

Visceral Fat and Exercise

Duke University Medical Center press release, "Physical Inactivity Rapidly Increases Visceral Fat; Exercise Can Reverse Accumulation," May 29, 2003, http://www.sciencedaily.com/releases/2003/05/030529081315.htm (accessed April 15, 2004).

Visceral Fat and Stress

Rob Stein, "New Theory Links Stress, 'Comfort Food' to Obesity," *San Jose Mercury News*, September 30, 2003 (report of "Chronic Stress and Obesity . . ." a paper published in the September 15, 2003, issue of the *Proceedings of the National Academy of Sciences* by Mary F. Dallman).

Underwood and Adler, "What You Don't Know about Fat," p. 45.

Health and Fat: It's Not What You Think

Gina Kolata, "What We Don't Know about Obesity," *New York Times*, June 22, 2003.

CHAPTER 5. WHO SAYS YOU'RE TOO FAT?

How We Got the BMI

Abby Ellin, "Quick, Do You Know Your B.M.I.?" *New York Times*, December 28, 2006.
Pond, *The Fats of Life*, p. 56.

Calculating BMI Using a Math Formula

"Body Mass Index Formula," Nutrition and Physical Activity pages, http://www.cdc.gov/nccdphp/dnpa/bmi/bmi-adult-formula.htm (accessed November 5, 2003).

Using a Web Site Calculator

See www.cdc.gov or www.nhlbisupport.com.

Interpreting the Results

"BMI Classification," World Health Organization: Global Database on Body Mass Index, December 7, 2006, http://www.who.int/bmi/index.jsp?introPage=intro_3.html (accessed July 10, 2007).
Ellin, "Quick, Do You Know Your B.M.I.?"

Maybe the Overweight Are Healthier

Paul Campos, *The Obesity Myth* (New York: Gotham Books, 2004), chap. 1.
Abby Ellin, "New Breed of Trainers Are Proving Fat Is Fit," *New York Times*, September 1, 2005.

Katherine M. Flegal et al., "Cause-Specific Deaths Associated with Under-weight, Overweight, and Obesity," *Journal of the American Medical Association* 298, no. 17 (November 7, 2007): 2028–37.

"Heavy People May Beat Critical Illnesses More Often," *New York Times*, May 9, 2006.

"Is It Healthier to Be a Little Overweight?" *Consumer Reports on Health*, October 2005, p. 8.

Gina Kolata, "A Matter of Fat," *AARP Bulletin*, June 2005, pp. 12–13.

———, "OK, Now We've Really Heard Everything," *San Jose Mercury News*, April 20, 2005 (report of new data from the Centers for Disease Control).

———, "Tell the Truth: Does This Index Make Me Look Fat?" *New York Times*, November 28, 2004.

James M. O'Brien Jr. et al., "Body Mass Index Is Independently Associated with Hospital Mortality in Mechanically Ventilated Adults with Acute Lung Injury," *Critical Care Medicine* 34, no. 3 (March 2006): 738–44.

Steven Shapin, "Eat and Run," *New Yorker*, January 16, 2006.

Waist-Hip Measurements as Health Predictors

"Is Your Heart at Risk? Get the Tape Measure," *New York Times*, November 15, 2005 (report of a study in the November 5, 2005 issue of *Lancet* led by Dr. Salim Yusuf).

"Past 75, Waist-Hip Ratio Trumps Height-Weight," *New York Times*, August 22, 2006, report of a study published in the August 2006 issue of the *American Journal of Clinical Nutrition*, Astrid Fletcher, senior author.

Salim Yusuf et al., "Obesity and the Risk of Myocardial Infarction in 27,000 Participants from 52 Countries: A Case-Control Study," *Lancet* 366, no. 9497 (November 5, 2005): 1640–49.

Maybe the Overweight Are Happier

Campos, *The Obesity Myth*, pp. 10–13.

Kenneth J. Mukamal et al., "Body Mass Index and Risk of Suicide among Men," *Archives of Internal Medicine* 167, no. 5 (March 12, 2007): 468–75.

"Suicide Found to Be Less Likely in Heavier People," *New York Times*, March 14, 2007 (report of a six-year study by Harvard University reported in the *Archives of Internal Medicine*).

An Alternative Set of Guidelines

Mary Enig and Sally Fallon, *Eat Fat Lose Fat* (New York: Hudson Street Press, 2005), pp. 106, 107.

Ways to Measure Body Fat

"Basal Metabolic Rate (BMR), Energy Expenditure, and Body Composition," Cornell natural science course 421 (undated), http://instructl.cit .cornell.edu.Courses/ns421/BMR.html (accessed November 19, 2003).

"BMI as a Reflection of the Body Energy Stores," Food and Agricultural Organization of the United Nations (undated), http://www.fao.org/ docrep/t1970e/t1970e03.htm (accessed November 17, 2003).

"Body Composition Tests" (explains dual-energy x-ray absorptiometry), http://www.topendsports.com (accessed November 14, 2003).

"Body Fat Measuring Device Aids in Weight Loss," September 12, 2003 (explanation of the Bod Pod), http://www.click2houston.com (accessed November 17, 2003).

"Body-Fat Scales—Will They Help?" *Consumer Reports*, January 2004, pp. 23–25.

Louisa Dalton, "Engineers Use Two Magnets Instead of One to Build a Lower-Cost MRI Scanner," *Stanford Report*, March 21, 2001.

"Measuring Body Fat," Obesity Resource Information Centre fact sheet (undated), http://ww.aso.org.uk/apps/oric/factsheets_content.asp (accessed November 19, 2003).

Norman Neill, "Tools to Measure Body Fat," Workshop Fitness, http://www .dinkyneill.hypermart.net (accessed November 19, 2003).

Other information came from the Web sites of body fat analyzers, such as bodytrends.com.

Pond, *The Fats of Life*, pp. 57–58.

Richard Seven, "Compose Yourself," *Pacific Northwest Magazine*, August 10, 2003.

E. L. Thomas et al., "Magnetic Resonance Imaging of Total Body Fat," *Journal of Applied Physiology* 85, no. 5 (November 1998): 1778–85.

Warren J. Willey II, "Measuring Fat—The 'Skinny' on Techniques" (undated), http://www.ironmagazine.com/archive/Measuring_Fat.htm (accessed November 19, 2003).

Yu Yevon, "The Ultimate Body Fat Testing Guide" (undated), http://www.abcbodybuilding.com/magazine/ultimatebodyfattestingguide.htm (accessed December 3, 2003).

Who's Fat and Who's Thin?

"Body Composition," Cornell natural science course 421, http://www.instructl.cit.cornell.edu (accessed November 17, 2003).

Wallace C. Donoghue, "The Importance of Body Composition and Percent Body Fat" (undated), http://www.bodytrends.com/buycalip.htm (accessed November 19, 2003).

Kolata, "A Matter of Fat."

Steven Shapin, "Eat and Run," *New Yorker*, January 16, 2006, pp. 79, 80.

P. S. Shetty and W. P. T. James, *Body Mass Index—A Measure of Chronic Energy Deficiency in Adults*, Food and Agricultural Organization of the United Nations Paper 56, chapter 2, "BMI as a Reflection of the Body Energy Stores," http://www.fao.org (accessed November 18, 2003).

CHAPTER 6. WHY YOUR BODY WANTS TO KEEP ITS SHAPE

Fighting Our Genes

Steven Greenhouse, "Overweight but Ready to Fight; Obese People Are Taking Their Bias Claims to Court," *New York Times*, April 4, 2003.

Gina Kolata, "How the Body Knows When to Gain or Lose," *New York Times*, September 17, 2000.

———, "Ideas and Trends: What We Don't Know about Obesity," *New York Times*, June 22, 2002.

Blame the Cave People

"Fighting the Thrifty Gene," *Scientific American Frontiers*, http://pbs.org/saf/1110/features/fighting.htm (accessed February 20, 2004).

James Heffley, "To Your Health," *Austin Chronicle*, December 14, 2001.

Ruppell, *The Hunger Gene*, pp. 162–66.

A Second Opinion

Taubes, *Good Calories, Bad Calories*, pp. 243–48.

Insulin Resistance

Jeffrey Friedman, "Genetics of Obesity and Type II Diabetes," *In the Lab*, Howard Hughes Medical Institute publication, July 12, 2002.

Gina Kolata, "Asking If Obesity Is a Disease or Just a Symptom," *New York Times*, April 16, 2002.

———, "Looking Past Blood Sugar Levels to Survive with Diabetes," *New York Times*, August 20, 2007, p. A1.

"Obesity Gene Pinpointed," BBC News, August 12, 2001, http://news.bbc.co.uk/1484659.stm (accessed November 13, 2003).

Meg Sullivan, "UCLA Study Finds Clues to Diabetes Puzzle," *Innovations Report*, May 6, 2003, http://www.innovations-report.com/html/reports/medicine_health/report-18969.html (accessed February 20, 2004).

Genetic Predisposition to Obesity

Victoria Stagg Elliott, "A Weighty Matter: Obesity, Leptin and Beyond," amednews.com, August 6, 2001, http://www.ama-assn.org/amednews/2001/08/06/hlsa0806.htm (accessed November 13, 2003).

Friedman, "Genetics of Obesity and Type II Diabetes."

Denise Grady, "Why We Eat (and Eat and Eat)," *New York Times*, November 16, 2002.

Robin Marantz Henig, "Fat Factors," *New York Times Magazine*, August 13, 2006, p. 30.

Matthew Hulver et al., "Gene Makes Muscles in the Obese Store More Fat," medicineworld.org, October 12, 2005 (report of study at Duke University), http://medicineworld.org/cancer/lead/11-2005/gene-that-makes-to-store-more-fat.html (accessed February 25, 2006).

Kolata, "Asking If Obesity Is a Disease or Just a Symptom."

———, "Genes Take Charge, and Diets Fall by the Wayside," *New York Times*, May 8, 2007, p. S1.

———, "How the Body Knows When to Gain or Lose."

"The Leptin Story in Obesity," *Duke Weight Loss Surgery Center: The Leptin Story* (undated), http://www.dukehealth.org/obesity/leptin_story.asp (accessed February 24, 2004).

Nicholas Wade, "Common Genetic Link to Obesity Is Discovered," *New York Times*, April 18, 2006 (report of study at Boston University).

———, "Genetic Cause Found for Some Cases of Human Obesity," *New York Times*, June 14, 1997.

How Those Fat Genes Work

Stuart Blackman, "The Hunger Hormone Unharnessed," *Scientist* 17, no. 19 (October 6, 2003): 30.

Didier Chapelot et al., "An Endocrine and Metabolic Definition of the Intermeal Interval in Humans: Evidence for a Role of Leptin on the Prandial Pattern through Fatty Acid Disposal," *American Journal of Clinical Nutrition* 72, no. 2 (August 2000): 421–31.

Jean Marx, "Cellular Warriors at the Battle of the Bulge," *Science* 299 (February 7, 2003): 846–49.

Pond, *The Fats of Life*, pp. 37, 299, 308.

Shell, *The Hungry Gene*, pp. 118–19.

Deborah A. Wilkinson, "A Weighty Matter: Neuropeptides Involved in Appetite and Energy Homeostasis," *Scientist* 13, no. 18 (September 13, 1999): 18.

The Role of Microorganisms

Henig, "Fat Factors," p. 31.

Peter Jaret, "Beating the Urge to Eat," *Reader's Digest*, July 2004, pp. 119–22.

Thomas H. Maugh II, "Bacteria Provide Clues on Obesity," *San Jose Mercury News*, December 21, 2006.

Wait—There's More

Glennda Chui, "Do These Genes Make Me Look Fat?" *San Jose Mercury News*, January 21, 2005.

Jeffrey M. Friedman, "A War on Obesity, Not the Obese," *Science* 299 (February 7, 2003): 856–58.

Nanci Hellmich, "Study: 'Puttering' Calories Count," *USA Today*, January 28, 2005 (refers to a study reported in the January 2005 issue of *Science*).

Jaret, "Beating the Urge to Eat."

Gina Kolata, "Find Yourself Packing It On? Blame Friends," *New York Times*, July 26, 2007, p. A1 (refers to a study reported in the July 26, 2007, issue of the *New England Journal of Medicine*).

———, "Why Some People Won't Be Fit Despite Exercise," *New York Times*, February 12, 2002.

Pond, *The Fats of Life*, p. 300.

Ruppel, *The Hungry Gene*, pp. 217–18.

Stein, "New Theory Links Stress, 'Comfort Food' to Obesity."

———, "Stress 'Trigger' with Diet Found as Key to Obesity," *San Jose Mercury News*, July 2, 2007, p. 6A.

A. Tremblay, J. P. Despres, and C. Bouchard, "Adipose Tissue Characteristics of Ex-obese Long-Distance Runners," *International Journal of Obesity* 8, no. 6 (1984): 641–48.

More Still to Come

Mary Carmichael, "A Changing Portrait of DNA," *Newsweek*, December 10, 2007, pp. 63–67.

Stephen S. Hall, "Small and Thin," *New Yorker*, November 19, 2007, pp. 55–57.

Peter Nathanielsz, *The Prenatal Prescription* (New York: Quill, 2001), pp. 47–51.

CHAPTER 7. CHOLESTEROL CONTROVERSIES

Introductory Paragraph

Daniel Carlat, "Dr. Drug Rep," *New York Times Magazine*, November 25, 2007.

Roy Moynihan and Alan Cassels, *Selling Sickness* (New York: Nation Books, 2005), p. 12.

Why Our Bodies Need Cholesterol

Enig, *Know Your Fats*, pp. 48–50, 56–58.

Beatric A. Golomb, Hakan Stattin, and Sarnoff Mednic, "Low Cholesterol and Violent Crime," *Journal of Psychiatric Research* 34 (July 2000): 301–309.

Rizwan M. Mufti, Richard Balond, and Cynthia L. Arfken, "Low Cholesterol and Violence," *Psychiatric Services* 49 (February 1998): 221–24.

Pond, *The Fats of Life*, p. 228.

Uffe Ravnskov, "Your Cholesterol Tells Very Little about Your Future Health," August 19, 2005, http://www.ravnskov.nu/ncep_guidelines (accessed August 19, 2005).

"Suicide Link to Cholesterol," BBC News report on an eight-year Finnish study, http://news.bbc.co.uk/1/hi/health/435305.stm (accessed December 12, 2006).

Jian Zhan, Matthew F. Muldoon, and Robert E. McKeown, "Serum Cholesterol Concentrations Are Associated with Visuomotor Speed in Men: Findings from the Third National Health and Nutrition Examination Survey, 1988–1994," *American Journal of Clinical Nutrition* 80, no. 2 (August 2004): 291–98.

What Are LDL and HDL?

Marcel Blanchaer, "Plasma Lipids," University of Manitoba Clinical Case Computer Tutorials in Biochemistry, September 28, 2002, http://www.umanitoba.ca/faculties/medicine/units/biochem/coursenotes/blanchaer_tutorial (accessed Setember 9, 2005).

Michael King, Medical Biochemistry at Indiana State (undated), http://www
.indstate.edu/thcme/mwking/lipoproteins.html (accessed March 22,
2004).

"Lipid Digestion and Lipoproteins," Medical Biochemistry, Indiana State
University, http://www.indstate.edu/thcme/mwking/llipoproteins.html,
(accessed March 22, 2004).

Pond, *The Fats of Life*, pp.123–24, 287–88.

Ravnskov, "Your Cholesterol Tells Very Little about Your Future Health."

Now, about Those Cholesterol Numbers

Howard Brody, *Hooked: Ethics, the Medical Profession, and the Pharmaceutical Industry* (Lanham, MD: Rowman and Littlefield, 2007), p. 33
(information about the expert panel of the National Cholesterol Education Project).

Mary Enig, "Why Is 5.2 (200) a 'Healthy' Cholesterol Level?" www.second
-opinions.com.uk./enig_chol.html (accessed August 18, 2005).

Gary Taubes, "The Soft Science of Dietary Fat," *Science* 291, no. 5513
(March 30, 2001): 2536–45 (in part 6).

High Cholesterol as a Predictor of Heart Disease

K. M. Anderson et al., "Cholesterol and Mortality: Thirty Years of Follow-
Up from the Framingham Study," *Journal of the American Medical
Association* 257, no. 16 (April 24, 1987): 2176–80.

H. M. Krumholz et al., "Lack of Association between Cholesterol and Coronary Heart Disease Mortality and Morbidity and All-Cause Mortality in
Persons Older Than 70 Years," *Journal of the American Medical Association* 272, no. 17 (November 2, 1994): 1335–40.

T. J. Moore, "The Cholesterol Myth," *Atlantic*, September 1989. (Moore, a
journalist, was an early skeptic of the relationship between diet and
blood cholesterol levels. I found his article on http://www.chelation
therapyonline.com/technical/p35.htm (accessed April 6, 2004), downloaded page 9.

Moynihan and Cassels, *Selling Sickness*, pp. 13, 14.

Uffe Ravnskov, "High Cholesterol May Protect against Infections and Atherosclerosis," *QJM* (an International Journal of Medicine) 96 (2003): 927–34.

Questioning the Role of Dietary Cholesterol in Heart Disease

Moore, "The Cholesterol Myth," downloaded p. 10.

Alice Ottoboni and Fred Ottoboni, "The Role of Cholesterol and Diet in Heart Disease," *21st Century Science and Technology* (Winter 2004–2005): 4–5.

Uffe Ravnskov, "Atherosclerosis and Coronary Heart Disease Have Nothing to Do with the Diet," January 1, 2001, http://www.ravnskov.nu/myth4 .htm (accessed May 25, 2005).

———, *The Cholesterol Myths*, pp. 111–12.

———, "New Cholesterol Guidelines for Converting Healthy People into Patients," October 31, 2003, http://www.ravnskov.nu/ncep_guidelines (accessed August 19, 2005).

Taubes, "The Soft Science of Dietary Fat," part 2.

What about Those Clogged Arteries?

"Atherosclerosis," http://en.wikipedia.org/wiki/Atherosclerosis (accessed April 4, 2007).

"Atherosclerosis and Greasy Sewer Lines," a discussion posted on the THINCS (The International Network of Cholesterol Skeptics) Web site, http://www.thincs.org/discuss.sewer.htm (accessed June 23, 2005).

C. V. Felton et al., "Dietary Polyunsaturated Fatty Acids and Composition of Human Aortic Plaques," *Lancet* 344, no. 8931 (October 29, 1994): 1195–96.

Atul Gawande, "The Way We Age Now," *New Yorker*, April 30, 2007, p. 50.

Gina Kolata, "It's Not a 'Plumbing Problem': Doctors Also Fight the Popular Misconceptions about the Causes of Heart Disease," *New York Times*, April 8, 2007.

———, "Scientists Begin to Question Benefit of 'Good' Cholesterol," *New York Times*, March 15, 2004.

Ann Underwood, "Quieting a Body's Defenses," *Newsweek Special Issue*, Summer 2005, p. 26.

Size Matters

Benoit Lamarche et al., "A Prospective Population-Based Study of Low Density Lipoprotein Particle Size as a Risk Factor for Ischemic Heart Disease in Men," *Canadian Journal of Cardiology* 17, no. 8 (August 17, 2001): 859–65.

Patty Siri and Ronald Krauss, "Influence of Dietary Carbohydrate and Fat on LDL and HDL Particle Distributions," *Current Atherosclerosis Reports* 7, no. 6 (November 2005): 455–59.

Taubes, *Good Calories, Bad Calories*, pp. 172–74.

Jeff Volek et al., "Modification of Lipoproteins by Very Low-Carbohydrate Diets," *Journal of Nutrition* 135, no. 6 (June 2005): 13389–42 (a review article of recent prospective studies comparing low-carbohydrate and high-carb/low-fat diets).

Effects of Cholesterol-Lowering Drugs

Sally Fallon and Mary G. Enig, "The Dangers of Statin Drugs: What You Haven't Been Told about Cholesterol-Lowering Medication," *Wise Traditions in Food, Farming, and the Healing Arts*, Weston A. Price Foundation, Spring 2004.

Ravnskov, "The Effect of the Statins Is Not Due to Cholesterol-Lowering," December 30, 2003, http://www.ravnskov.nu/muth6.htm (accessed May 25, 2005).

———, "New Cholesterol Guidelines for Converting Healthy People into Patients."

Side Effects of Cholesterol-Lowering Drugs

"Cholesterol Level and Parkinson's May Be Linked," *New York Times*, January 2, 2007 (report of a study published December 18, 2006, in the journal *Movement Disorders*).

Fallon and Enig, "The Dangers of Statin Drugs: What You Haven't Been Told about Cholesterol-Lowering Medication."

Moynihan and Cassels, *Selling Sickness*, pp. 17–19.

"The Statin Effects Study," UCSD Statin Effects Study Newsletter, May 17,

2007 (report of University of California at San Diego collection of statin side effects as reported by 4112 people; undated), http://medicine.ucsd .edu/ses/adverse_effects.htm (accessed March 15, 2007).

An Epidemic of Diagnoses

H. Gilbert Welch, Lisa Schwartz, and Steven Woloshin, "What's Making Us Sick Is an Epidemic of Diagnoses," *New York Times*, January 2, 2007.

CHAPTER 8. SATURATED FATS: HEALTHFUL FOOD

Fat Mixtures: Saturated with Unsaturated

Enig, *Know Your Fats*, pp. 17–18, 30–31.
Nina Planck, *Real Food* (New York: Bloomsbury, 2004), pp. 107–108, 168–70.
Pond, *The Fats of Life*, pp. 79–81.
Andrew L. Stoll, *The Omega-3 Connection* (New York: Fireside, 2001) p. 33.
Taubes, "The Soft Science of Saturated Fat," part 6.

Mixed-Up Molecules

Enig, *Know Your Fats*, pp. 30–31.
J. Bruce German and Cora J. Dillard, "Saturated Fats: What Dietary Intake?" *American Journal of Clinical Nutrition* 80, no. 3 (September 2004): 550–59.
Antonio Zamora, "Fats, Oils, Fatty Acids, Triglycerides" (undated), subsection "Fatty Acid Composition of Some Common Edible Fats and Oils," http://www.scientificpsychic.com/fitness/fattyacids.html (accessed April 10, 2007).

What about Cholesterol?

Mary Enig and Sally Fallon, "The Skinny on Fats," in *Nourishing Traditions* (Winona Lake, IN: New Trends Publishing, 1999). http://www.westona

price.org/knowyourfats/skinny, downloaded pages 1–11 (accessed July 8, 2007).

German and Dillard, "Saturated Fats: What Dietary Intake?"

J. H. Hays et al., "Effect of a High Saturated Fat and No-Starch Diet on Serum Lipid Subfractions in Patients with Documented Atherosclerotic Cardiovascular Disease," *Mayo Clinic Proceedings* 78, no. 11 (November 2003): 1331–36.

Saturated Fat and Heart Disease

Enig, "The Skinny on Fats," downloaded pages 2–6.

German and Dillard, "Saturated Fats: What Dietary Intake?"

Ravnskov, *The Cholesterol Myths*, pp. 16–48.

Taubes, "The Soft Science of Saturated Fat," part 7.

A Persistent Myth

A. Ascherio et al., "Dietary Fat and Risk of Coronary Heart Disease in Men: Cohort Follow-Up Study in the United States," *British Medical Journal* 313 (July 1996): 84–90.

L. A. Corr and M. F. Oliver, "The Low-Fat, Low-Cholesterol Diet Is Ineffective," *European Heart Journal* 18 (1997): 18–22.

German and Dillard, "Saturated Fats: What Dietary Intake?"

Ravnskov, *The Cholesterol Myths*, p. 5.

S. L. Weinbeg, "The Diet-Heart Hypothesis: A Critique," *Journal of the American College of Cardiology* 44, no. 9 (March 2004): 725–30.

Nutritional Benefits of Saturated Fats

Mary G. Enig, "The Importance of Saturated Fats for Biological Functions," http://www.westonaprice.org/knowyourfats/import_sat_fat.html (accessed February 27, 2007).

———, "The Importance of Saturated Fats for Biological Functions," *Wise Traditions in Food, Farming, and the Healing Arts*, Spring 2004.

German and Dillard, "Saturated Fats: What Dietary Intake?"

Interview of Bruce Fife by Rachael Basely, "Things You Probably Didn't

Know about Coconut Oil," *Doctors' Prescription for Healthy Living* 7, no. 6 (2004): 48–49.

"Medium-Chain Triglycerides," PDR health, www.pdrhealth.com/drug_/info/ nmdrugprofiles/nutsupdrugs/med_0172.shtml (accessed March 27, 2004).

Planck, *Real Food*, pp. 174–79, 190–94.

CHAPTER 9: OILS: ESSENTIAL AND OTHERWISE

Monounsaturated Oils

Sally Fallon and Mary G. Enig, "The Great Con-ola," http://www.westona price.org/knowyourfats/conola.html (accessed June 13, 2007).

"Oil and Vinegar," *Wine Spectator*, September 30, 2005, p. 176.

"Scientists Shed Light on a Secret of the Olive Tree," *New York Times*, September 6, 2005.

"Virgin Olive Oil Helps Improve Your Circulation," *Consumer Report on Health*, March 2006.

Polyunsaturated Oils

W. W. Christie, "Fatty Acids: Methylene-Interrupted Double Bonds," http:// www.lipidlibrary.co/uk (accessed January 23, 2007).

Enig, *Know Your Fats*, pp. 37–38.

What Does "Omega" Mean?

Enig, *Know Your Fats*, p. 25.

What's Essential?

"Essential Fatty Acid," http://en.wikipedia.org/wiki/Essential_fatty_acid (accessed January 21, 2007).

Planck, *Real Food*, p. 124.

Andrew L. Stoll, *The Omega-3 Connection* (New York: Fireside Books, 2001), p. 33.

Essential Oils in Meats and Dairy Products

Enig, *Know Your Fats*, p. 46.

Annika Smedman et al., "Pentadecanoic Acid in Serum as a Marker for Intake of Milk Fat: Relations between Intake of Milk Fat and Metabolic Risk Factors," *American Journal of Clinical Nutrition* 69, no. 1 (January 1999): 22–29.

Why All the Fuss about Omega-3 Oil?

"The Claim: Eating Fish Is Good for the Brain," *New York Times*, January 3, 2006.

"Fish Oil for Mom May Benefit Her Child," *New York Times*, January 2, 2007.

"Omega-3 Fatty Acid," http://en.wikipedia.org/wiki/Omega-3_fatty_acid (accessed January 26, 2007).

Stoll, *The Omega-3 Connection*, pp. 15, 22, 37, 67–75.

Dangers of Polyunsaturated Fats

Christie, "Fatty Acids: Methylene-Interrupted Double Bonds."

Enig and Fallon, "The Skinny on Fats," downloaded pages 8, 9.

Stoll, *The Omega-3 Connection*, p. 201.

Choosing Oils

Enig, *Know Your Fats*, pp. 18–19, 148.

Olive Oil Terminology

Tom Mueller, "Slippery Business," *New Yorker*, August 13, 2007, pp. 38–45.

"Olive Oil Definitions," The Olive Oil Source, http://www.oliveoilsource .com/definitions.htm (accessed July 5, 2007).

How Cooking Affects Fats

Enig, *Know Your Fats*, p. 197.
Pond, *The Fats of Life*, p. 107.

Storing Fats and Oils

Pond, *The Fats of Life*, pp. 105–106.

CHAPTER 10. WHAT'S WRONG WITH TRANS FATS

How Trans Fats Are Made

Pond, *The Fats of Life*, p. 100.
Robert L. Wolke, "Inside the New Crisco: Losing Trans Fatty Acids," *San Jose Mercury News*, April 26, 2006.
Antonio Zamora, "Fats, Oils, Fatty Acids, Triglycerides" (undated), subsection "What Is Hydrogenation and Partial Hydrogenation?" http://www.scientificpsychic.com/fitness/fattyacids.html (accessed April 10, 2007).

Trans Fats in Our Diets

Ravnskov, *The Cholesterol Myths*, p. 237.
Kim Severson, *The Trans Fat Solution* (Berkeley, CA: Ten Speed Press, 2003), pp. 1–7.

The Problem with Human-Made Trans Fats

Comments to FDA in response to the Advance Notice of Proposed Rulemaking, submitted October 9, 2003 (bantransfats.com), a California nonprofit corporation.
Enig and Fallon, "The Skinny on Fats" (under "Hydrogenation," downloaded page 13).
Nicholas D. Kristof, "Killer Girl Scouts," *New York Times*, May 21, 2006.
Fred A. Kummerow et al., "Effect of Trans Fatty Acids on Calcium Influx

into Human Arterial Cells," *American Journal of Clinical Nutrition* 70, no. 5 (November 1999): 832–38.

Severson, *The Trans Fat Solution*, pp. 8–10.

Nina Teicholz, "Nuggets of Death," *New York Times*, April 16, 2006.

Trans Fats at the Market

Jeffrey Kluger, "Fessing Up to Fats," *Time Canada*, July 21, 2003.

Kim Severson and Melanie Warner, "Fat Substitute, Once Praised, Is Pushed Out of the Kitchen," *New York Times*, July 13, 2005.

Wolke, "Inside the New Crisco: Losing Trans Fatty Acids."

Trans Fats at the Restaurant

Linda A. Johnson, "Fat Content of Fast Food Varies Widely, Study Finds," *San Jose Mercury News*, April 13, 2006.

Thomas J. Lueck and Kim Severson, "New York Bans Most Trans Fats in Restaurants," *New York Times*, December 6, 2006.

Michael Mason, "A Dangerous Fat and Its Risky Alternatives," *New York Times*, October 10, 2006.

Severson and Warner, "Fat Substitute, Once Praised, Is Pushed Out of the Kitchen."

How to Identify Trans Fats in Food Labels

Severson, *The Trans Fat Solution*, pp. 10–13, 15–16.

Natural Trans Fats

"Naturally Occurring Trans Fats," tfX.org, http://www.tfx.org.uk/page62 .html (accessed March 12, 2007).

Janet Raloff, "Inflammation-Fighting Fat," *Science News Online*, October 29, 2005, http://www.sciencenews.org/scripts (accessed March 9, 2006).

Lumping the Good with the Bad

Kim Severson, "Trans Fat Fight Claims Butter as a Victim," *New York Times*, March 7, 2007.

And Now, a New Trans Fat Substitute

Ben Harder, "A Trans Fat Substitute Might Have Health Risks Too," *Science News Online*, week of February 10, 2007, http://www.sciencenews.org/scripts (accessed February 26, 2007).
Corby Kummer, "High on the Hog," *New York Times*, August 12, 2005.

CHAPTER 11. THE PROBLEM WITH LOW-FAT DIETS

Where's the Beef?

Taubes, "The Soft Science of Dietary Fat," part 1.

Some History

Enig and Fallon, *Eat Fat Lose Fat*, pp. 23, 35.
Taubes, "The Soft Science of Dietary Fat," part 2.

Faulty Food Pyramids

Alice Ottoboni and Fred Ottoboni, "The Food Guide Pyramid: Will the Defects Be Corrected?" *Journal of American Physicians and Surgeons* 9, no. 4 (2004): 109–13.
Gary Taubes, "What If It's All Been a Big Fat Lie?" *New York Times Magazine*, July 7, 2002, pp. 24–25.
Walter C. Willett and Meir J. Stampler, "Rebuilding the Food Pyramid," *Scientific American*, December 17, 2002, http://www.sciam.com (accessed April 27, 2005).

The Latest Pyramid

Loren Cordain, "Cereal Grains: Humanity's Double-Edged Sword," in *Evolutionary Aspects of Nutrition and Health: Diet, Exercise, Genetics and Chronic Disease*, ed. A. P. Simopoulos (Basel: Karger, 1999), pp. 19–73.

"Food Pyramids," *Nutrition Source, Harvard School of Public Health*, http://www.hsph.harvard.edu/nutritionsource/pyramids.html (accessed April 27, 2005).

Kim Severson, "The Government's Pyramid Scheme," *New York Times*, April 24, 2005.

———, "When a Food Marketer Helps Devise Nutrition Advice," *New York Times*, April 16, 2005, p. A18.

Aleta Watson, "Food Pyramid Upheaval Aims to Customize Diets," *San Jose Mercury News*, April 20, 2005, p. 3A.

The Problem with Refined Carbohydrates

John R. Guyton, Eric C. Westman, and William S. Yancy Jr., "Clinical Insights #1: Low Carbohydrate Diets for the Treatment of Obesity and Hypertriglyceridemia Division of General Internal Medicine," Duke University Medical Center, http://www.lipid.org/clinical/insights/1000001.php (accessed April 8, 2005).

"Insulin Resistance," http://syndromex.stanford.edu/InsulinResistance.htm (accessed April 12, 2005).

Elizabeth J. Parks and Marc K. Hellerstein, "Carbohydrate-Induced Hypertriacyglycerolemia: Historical Perspectives and Review of Biological Mechanisms," *American Journal of Clinical Nutrition* 71 (2000): 412–33.

Michael F. Roizen and Mehmet C. Oz, *You on a Diet—The Owner's Manual for Waist Management* (New York: Free Press, 2006), p. 67.

Taubes, *Good Calories, Bad Calories*, pp. 385, 389–90.

———, "What If It's All Been a Big Fat Lie?" p. 27.

The Importance of Glycemic Levels

Enig and Fallon, *Eat Fat Lose Fat*, p. 69.

Eric Oliver, *Fat Politics—The Real Story behind America's Obesity Epidemic* (New York: Oxford University Press, 2006), p. 117.

Taubes, "What If It's All Been a Big Fat Lie?" p. 27.

Low-Fat Milk—Better Think Twice

J. E. Chavarro et al., "A Prospective Study of Dairy Foods Intake and Anovulatory Infertiliy," *Human Reproduction* 22 (February 28, 2007): 1340–47.

Sally Fallon, "Dirty Secrets of the Food Processing Industry" (a presentation at the annual conference of Consumer Health of Canada, March 2002), http://www.westonaprice.org/modernfood/dirty-secrets.html (accessed July 5, 2007).

Planck, *Real Food*, p. 66.

Low-Fat Diets and Mental Health

J. R. Kaplan et al., "The Effects of Fat and Cholesterol on Social Behavior in Monkeys," *Psychosomatic Medicine* 53 (November–December 1991): 634–42.

Pond, *The Fats of Life*, pp. 227–28.

A. S. Wells et al., "Alterations in Mood after Changing to a Low-Fat Diet," *British Journal of Nutrition* 79 (January 1998): 23–30.

Weight-Loss Diets Compared

Christopher D. Gardner et al., "Comparison of the Atkins, Zone, Ornish, and LEARN Diets for Change in Weight and Related Risk Factors among Overweight Premenopausal Women—The A to Z Weight Loss Study: A Randomized Clinical Trial," *Journal of the American Medical Association* 297 (March 7, 2007): 969–77.

Barbara Feder Ostrov, "Study Throws Weight behind Low-Carb Diets," *San Jose Mercury News*, March 7, 2007, p. 1A.

Low-Fat Diets Laid to Rest—I Wish

Gina Kolata, "Low-Fat Diet Does Not Cut Health Risks, Report Says," *New York Times*, February 8, 2006, p. A1.

R. L. Prentice et al., "Low-Fat Dietary Pattern and Risk of Invasive Breast Cancer: The Women's Health Initiative Randomized Controlled Dietary Modification Trial," *Journal of the American Medical Association* 295 (February 8, 2007): 655–66.

CHAPTER 12. DIVERSITY AND BALANCE

Variations in Diet/Gene Responses: Some Examples

Jan L. Breslow, description of current research of laboratory heads, Rockefeller University, http://www.rockefeller.edu/research/abstract.php?id +126 (accessed February 21, 2007).

Caroline F. Bunn et al., "Mutation in Brief," *Human Mutation* 492 (2001), http://www.interscience.wiley.com/humanmutation/pdf/mutation/492.pdf (accessed July 9, 2007).

"Familial hypercholesterolemia," http://en.wikipedia.org/wiki/Familial _hypercholesterolemia (accessed July 8, 2007).

R. A. Hegele et al., "Genetic Variation of Intestinal Fatty Acid-Binding Protein Associated with Variation in Body Mass in Aboriginal Canadians," *Journal of Clinical Endocrinology and Metabolism* 81 (1996): 4334–37.

M. K. Hofman and H. M. G. Princen, "Project: Genetic Variation in Bile Acid Metabolism and the Response to Dietary Fat" (doctoral dissertation), April 2005, www.onderzoekinformatie.nl/en/oi/nod/onderzoek/ OND1277781 (accessed February 22, 2007).

Jose M. Ordovas, "Dietary Fat Intake Determines the Effect of a Common Polymorphism in the Hepatic Lipase Gene Promoter on High-Density Lipoprotein Metabolism," *Circulation* online, http://www.circ.aha journals.org/cgi/content/full/106/18/2315 (accessed February 26, 2007).

Nutritional Genomics

Carren Bersch, "Nutritional Genomics: Part of the Evolution of Personalized Medicine," *Medical Laboratory Observer*, July 2006, http://www .findarticles.com (accessed February 22, 2007).

Bruce Grierson, "What Your Genes Want You to Eat," *New York Times Magazine*, May 4, 2003.

Corby Kummer, "Your Genomic Diet," *Technology Review* online, August 2005, www.technologyreview.com (accessed February 22, 2007).

Peg Tyre, "You Are What You Eat," *Newsweek*, October 31, 2005, p. 63.

A Balance of Fats

K. C. Hayes, "Dietary Fat and Heart Health: In Search of the Ideal Fat," *Asia Pacific Journal of Clinical Nutrition* 11 (supplement) (2002): S394–400.

The Importance of Balancing Omega-6 and Omega-3 Oils

A. P. Simopoulos, "The Importance of the Ratio of Omega-6/Omega-3 Essential Fatty Acids," *Biomedicine Pharmacotherapy* 56, no. 8 (October 2002): 365–79.

Stoll, *The Omega-3 Connection*, pp. 38–43.

The Consequences of an Omega-6 and Omega-3 Imbalance

Enig and Fallon, "The Skinny on Fats," downloaded page 9.

Stoll, *The Omega-3 Connection*, pp. 43–47, 180–200.

Why the Imbalance

Stoll, *The Omega-3 Connection*, pp. 43–50, 56.

Striking the Essential Oil Balance

Mary Enig, *Know Your Fats*, pp. 104–108.

Our Balancing Mechanisms

Taubes, *Good Calories, Bad Calories*, pp. 295–99, 390, 439.

CHAPTER 13. FAT, THE FARM, AND FAMILY

Rivers of Corn Syrup

Michael Pollan, *The Omnivore's Dilemma—A Natural History of Four Meals* (New York: Penguin Press, 2006), p. 104.

Arnold H. Slyper, "The Pediatric Obesity Epidemic: Causes and Controversies," *Journal of Clinical Endocrinology and Metabolism* 89, no. 6 (2004): 2540–47.

Mountains of Corn

Pollan, *The Omnivore's Dilemma*, pp. 104, 115–16.

Spoonfuls of Broccoli

Marian Burros, "The Debate over Subsidizing Snacks," *New York Times*, July 4, 2007, p. D1.

"Cut the Fat," *Consumers Report*, January 2004, p. 15.

David M. Herszenhorn, "Farm Subsidies Seem Immune to an Overhaul," *New York Times*, July 26, 2007, p. A1.

Solo Snacking

Jon Mooallem, "Twelve Easy Pieces," *New York Times Magazine*, February 12, 2006, pp. 45–46.

Oliver, *Fat Politics*, pp. 133–34.

Jeremy W. Peters, "Fewer Bites. Fewer Calories. Lot More Profit." *New York Times*, July 7, 2007, p. B1.

Ounces of Leaves

Michael Pollan, "Unhappy Meals," *New York Times Magazine*, January 28, 2007, p. 47.

Anne Underwood, "Mmmm, Tasty Chemicals," *Newsweek*, March 5, 2007, p. 50.

Toddlers on the Bandwagon

T. A. Badger, "Poor Nutrition Habits Start Early," *San Jose Mercury News*, 5A (report of the "Feeding Infants and Toddlers Study" of 2002).

Carla K. Johnson, "Researchers: Think Twice before Giving Preschoolers Juice," *San Jose Mercury News*, February 7, 2005.

Poorer yet Fatter

Michael Pollan, "You Are What You Grow," *New York Times Magazine*, April 22, 2007, pp. 15–16.

Elizabeth Weil, "Heavy Questions," *New York Times Magazine*, January 2, 2005.

Intervening in the Schools

Benjamin Caballero et al., "Pathways: A School-Based, Randomized Controlled Trial for the Prevention of Obesity in American Indian Schoolchildren," *American Journal of Clinical Nutrition* 78, no. 5 (November 2003): 1030–38.

Gina Kolata, *Rethinking Thin: The New Science of Weight Loss—and the Myths and Realities of Dieting* (New York: Farrar, Straus and Giroux, 2007), pp. 197–200.

Philip R. Nader et al., "Three-Year Maintenance of Improved Diet and Physical Activity—the CATCH Cohort," *Archives of Pediatrics and Adolescent Medicine* 153, no. 7 (July 1999): 695–704.

Programming Babies

Annie Murphy Paul, "Baby's First Diet Pill," *New York Times Magazine*, August 5, 2007.

Old-Fashioned Eating

Jack Challem, "Paleolithic Nutrition: Your Future Is in Your Dietary Past," http://nutritionreporter.com/stone_age_diet.html (accessed November 7, 2003).

Stanley M. Garn and William R. Leonard, "What Did Our Ancestors Eat?" http://naturalhygienesociety.org/articles/paleo1/html (accessed June 4, 2006).

Finding Real Food

Sally Fallon and Mary G. Enig, "Splendor from the Grass," *Wise Traditions in Food, Farming, and the Healing Arts*, Summer 2000, http://www.westonaprice.org/farming/splendor.html (accessed February 27, 2007).

Pollan, *The Omnivore's Dilemma*, pp. 77–79, 267.

An Evolution Solution

Nicholas Wade, "Humans Have Spread Globally, and Evolved Locally," *New York Times*, June 26, 2007, p. D3.

BIBLIOGRAPHY

Note: For periodicals, see "References."

Atkins, Robert C. *Dr. Atkins' New Diet Revolution*. New York: Avon Books, 1992.

Brody, Howard. *Hooked: Ethics, the Medical Profession, and the Pharmaceutical Industry*. Lanham, MD: Rowman and Littlefield, 2007.

Campos, Paul. *The Obesity Myth: Why America's Obsession with Weight Is Hazardous to Your Health*. New York: Gotham Books, 2004.

Enig, Mary G. *Know Your Fats: The Complete Primer for Understanding the Nutrition of Fats, Oils, and Cholesterol*. Silver Spring, MD: Bethesda Press, 2000.

Enig, Mary G., and Sally Fallon. *Eat Fat Lose Fat*. New York: Hudson Street Press, 2005.

Gershon, Michael D. *The Second Brain*. New York: HarperPerennial, 1998.

Guiliano, Mireille. *French Women Don't Get Fat: The Secret of Eating for Pleasure*. New York: Alfred A. Knopf, 2005.

Kolata, Gina. *Rethinking Thin: The New Science of Weight Loss—and the Myths and Realities of Dieting*. New York: Farrar, Straus and Giroux, 2007.

Leith, William. *The Hungry Years: Confessions of a Food Addict*. New York: Gotham Books, 2005.

Moynihan, Ray, and Alan Cassels. *Selling Sickness: How the World's Biggest Pharmaceutical Companies Are Turning Us All into Patients*. New York: Nation Books, 2005.

Nathanielsz, Peter. *The Prenatal Prescription*. New York: Quill, 2001.

Nuland, Sherwin B. *How We Live*. New York: Vintage Books, 1998.

Oliver, J. Eric. *Fat Politics: The Real Story behind America's Obesity Epidemic*. New York: Oxford University Press, 2006.

Planck, Nina. *Real Food: What to Eat and Why*. New York: Bloomsbury, 2006.

Pollan, Michael. *The Omnivore's Dilemma: A Natural History of Four Meals*. New York: Penguin, 2006.

Pond, Caroline M. *The Fats of Life*. Cambridge: Cambridge University Press, 1998.

Ravnskov, Uffe. *The Cholesterol Myths: Exposing the Fallacy That Saturated Fat and Cholesterol Cause Heart Disease*. Washington, DC: New Trends, 2000.

Roizen, Michael F., and Mehmet C. Oz. *You on a Diet: The Owner's Manual for Waist Management*. New York: Free Press, 2006.

Russell, Sharman Apt. *Hunger: An Unnatural History*. New York: Basic Books, 2005.

Sears, Barry. *The Zone*. New York: Regan Books, 1995.

Severson, Kim. *The Trans Fat Solution: Cooking and Shopping to Eliminate the Deadliest Fat from Your Diet*. Berkeley, CA: Ten Speed Press, 2003.

Shell, Ellen Ruppel. *The Hungry Gene: The Science of Fat and the Future of Thin*. New York: Atlantic Monthly Press, 2002.

Stoll, Andrew L. *The Omega-3 Connection*. New York: Fireside, 2001.

Taubes, Gary. *Good Calories, Bad Calories*. New York: Alfred A Knopf, 2007.

Weil, Andrew. *Healthy Aging: A Lifelong Guide to Your Physical and Spiritual Well-Being*. New York: Alfred A. Knopf, 2005.

Willitt, Walter C. *Eat, Drink, and Be Healthy: The Harvard Medical School Guide to Healthy Eating*. New York: Free Press, 2001.

INDEX